Solutions Guide

Introductory Chemistry:
A Foundation
Introductory Chemistry
Basic Chemistry

Fifth Edition

Zumdahl

James F. Hall
University of Massachusetts Lowell

Houghton Mifflin Company
Boston   **New York**

Vice President and Publisher: Charles Hartford
Executive Editor: Richard Stratton
Senior Development Editor: Bess Deck
Editorial Assistant: Rosemary Mack
Senior Project Editor: Cathy Brooks
Senior Manufacturing Coordinator: Priscilla Bailey
Senior Marketing Manager: Katherine Greig

Printed in the U.S.A.

ISBN: 0-618-30530-0

3456789 – BBS – 07 06 05 04

Table of Contents

Preface v

Chapter 1   Chemistry: An Introduction 1

Chapter 2   Measurements and Calculations 2

Chapter 3   Matter and Energy..................................... 15

Cumulative Review: Chapters 1, 2, and 3 21

Chapter 4   Chemical Foundations: Elements, Atoms, and Ions........ 26

Chapter 5   Nomenclature.. 34

Cumulative Review: Chapters 4 and 5................................. 42

Chapter 6:   Chemical Reactions: An Introduction 47

Chapter 7:   Reactions in Aqueous Solutions........................ 52

Cumulative Review: Chapters 6 and 7 63

Chapter 8:   Chemical Composition 68

Chapter 9:   Chemical Quantities 94

Cumulative Review: Chapters 8 and 9 115

Chapter 10:   Modern Atomic Theory 121

Chapter 11:   Chemical Bonding 128

Cumulative Review: Chapters 10 and 11 142

Chapter 12:   Gases 148

Chapter 13:   Liquids and Solids 165

Chapter 14:   Solutions 171

Cumulative Review: Chapters 12, 13, and 14 188

Chapter 15:   Acids and Bases 196

Chapter 16:   Equilibrium 206

Cumulative Review: Chapters 15 and 16 215

Chapter 17:   Oxidation-Reduction Reactions/Electrochemistry 221

Chapter 18:   Radioactivity and Nuclear Energy 233

Chapter 19:   Organic Chemistry 239

Chapter 20:   Biochemistry 251

Preface

This guide contains detailed solutions for the even-numbered end-of-chapter problems in the fifth editions of *Introductory Chemistry*, *Introductory Chemistry: A Foundation*, and *Basic Chemistry* by Steven S. Zumdahl. Hundreds of new problems and questions have been prepared for the new editions of the text, which we hope will be of help to you in gaining an understanding of the fundamental principles of chemistry.

Think of this guide as the last chapter of a mystery novel. Just as you wouldn't initially turn to the last few pages of a mystery to find out "whodunit," you should not look at the solutions in this guide until you have tried diligently to solve the end-of-chapter problems yourself. To learn chemistry effectively, you have to spend a lot of time pushing a pencil around on paper!

As you study chemistry, first go over your lecture notes and read through your textbook to see if there is any factual material or terminology you don't understand. Then work through the example problems your professor had done in class and the example problems and self-check exercises in the textbook, paying close attention to the *method of solution* used for each problem. Then try again to work through the same classroom and textbook examples yourself without looking at the solutions. Once you feel comfortable with the example problems, you can feel confident to tackle the end-of-chapter problems. After you've mastered a few chapters, then test your understanding by going through the Cumulative Review material provided for those chapters. These reviews tie together the material covered in the previous few chapters and are intended to give you some perspective on your progress in chemistry.

One topic that causes students much concern is the matter of significant figures, and the determination of the number of digits to which a solution to a problem should be reported. To avoid truncation errors in the solutions contained in this guide, the solutions typically report *intermediate* answers to one more digit that appropriate for the *final* answer. The final answer to each problem is then given to the correct number of significant figures based on the data provided in the problem. Realize especially that your calculator doesn't know anything about the rules for significant figures: a calculator will report answers to however many digits it is programmed for, regardless of whether those digits mean anything. You have to have the understanding of the measurements involved and have to do the rounding of answers yourself.

Many very dedicated people have worked long hours to prepare this guide. Particular thanks go to Rosemary Mack, Editorial Assistant at Houghton Mifflin Company, for her patience, competence, and good cheer.

Good luck in your study of chemistry!

James F. Hall
James_Hall@uml.edu

Solutions Guide

Introductory Chemistry:
A Foundation
Introductory Chemistry
Basic Chemistry

Chapter 1   Chemistry: An Introduction

2.   The answer will depend on student examples.

4.   There are, unfortunately, many examples. Chemical and biological weapons are still being produced in some countries. Although the development of new plastics has been a boon in many endeavors, this also increases our depletion of fossil fuels and our solid waste problems. Although many exciting new drugs and treatments have become available, the same biotechnology may lead to testing procedures for determining whether a person has a genetic likelihood of developing a particular disease, which may make it impossible or difficult for that person to obtain health or life insurance.

6.   This answer depends on your own experience, but consider the following examples: oven cleaner (the label says it contains sodium hydroxide; it converts the burned-on grease in the oven to a soapy material that washes away); drain cleaner (the label says it contains sodium hydroxide; it dissolves the clog of hair in the drain); stomach antacid (the label says it contains calcium carbonate; it makes me belch and makes my stomach feel better); hydrogen peroxide (the label says it is a 3% solution of hydrogen peroxide; when applied to a wound, it bubbles); depilatory cream (the label says it contains sodium hydroxide; it removes unwanted hair from skin).

8.   The scientist must recognize the problem and state it clearly, propose possible solutions or explanations, and then decide through experimentation which solution or explanation is best.

10.   a.   quantitative - a number (measurement) is indicated explicitly
       b.   qualitative - a color is a qualitative description
       c.   qualitative - no measurement is implied
       d.   quantitative - the size of the hailstones is a measurement
       e.   qualitative - no measurement is implied
       f.   quantitative - the distance between the cities is a measurement

12.   A natural law is a *summary of observed, measurable behavior* that occurs repeatedly and consistently. A theory is our attempt to *explain* such behavior. The conservation of mass observed during chemical reactions is an example of a natural law. The idea that the universe began with a "big bang" is an example of a theory.

14.   Most applications of chemistry are oriented toward the interpretation of observations and the solving of problems. Although memorization of some facts may *aid* in these endeavors, it is the ability to combine, relate, and synthesize information that is most important in the study of chemistry.

16.   In real life situations, the problems and applications likely to be encountered are not simple textbook examples. One must be able to observe an event, hypothesize a cause, and then test this hypothesis. One must be able to carry what has been learned in class forward to new, different situations.

Chapter 2   Measurements and Calculations

2.   Any number can be expressed as the product of a number between 1 and 10
      and a power of 10 (either positive or negative). The power of 10 depends
      on the number of places to the left (positive) or right (negative) that
      the decimal point is moved.

4.   positive; negative

6.   a.   positive

      b.   positive

      c.   negative

      d.   negative

8.   a.   The decimal point must be moved six places to the left, so the ex-
            ponent is positive 6; $9,367,421 = 9.367421 \times 10^6$.

      b.   The decimal point must be moved three places to the left, so the
            exponent is positive 3; $7241 = 7.241 \times 10^3$.

      c.   The decimal point must be moved four places to the right, so the
            exponent is negative 4; $0.0005519 = 5.519 \times 10^{-4}$.

      d.   The decimal point does not have to be moved, so the exponent is
            zero; $5.408 = 5.408 \times 10^0$.

      e. 6.24×10^2 is already written in standard scientific notation.

      f.   The decimal point must be moved three places to the left, and the
            resulting exponent of positive three must be combined with the ex-
            ponent of negative two in the multiplier; $6,319 \times 10^{-2} = 6.319 \times 10^1$.

      g.   The decimal point must be moved nine places to the right, so the
            exponent is negative nine; $0.000000007215 = 7.215 \times 10^{-9}$

      h.   The decimal point must be moved one place to the right, so the ex-
            ponent is negative 1; $0.721 = 7.21 \times 10^{-1}$

10.   a.   The decimal point must be moved four places to the left;
            $7.327 \times 10^{-4} = 0.0007327$.

      b.   The decimal point must be moved two places to the right;
            $1.51 \times 10^2 = 151$.

      c.   The decimal point does not have to be moved (zero places);
            $1 \times 10^0 = 1$.

      d.   The decimal point must be moved four places to the left;
            $5.399 \times 10^{-4} = 0.0005399$.

e.　The decimal point must be moved three places to the right;
0.221 × 10³ = 221.

f.　The decimal point must be moved two places to the left;
7.83 × 10⁻² = 0.0783.

g.　The decimal point must be moved four places to the left;
1218 × 10⁻⁴ = 0.1218.

h.　The decimal point must be moved four places to the left;
2.918 × 10⁻⁴ = 0.0002918.

i.　The decimal point must be moved three places to the right;
7.251 × 10³ = 7251.

j.　The decimal point must be moved nine spaces to the left;
1.911 × 10⁻⁹ = 0.000000001911.

k.　The decimal point must be moved two spaces to the right;
9.951 × 10² = 995.1.

l.　The decimal point must be moved two spaces to the left;
9.951 × 10⁻² = 0.09951.

12.　To say that scientific notation is in *standard* form means that you have a number between 1 and 10, followed by an exponential term. The numbers given in this problem are *not* between 1 and 10 as written.

a.　$8714 \times 10^2 = (8.714 \times 10^3) \times 10^2 = 8.714 \times 10^5$

b.　$0.0006591 \times 10^8 = (6.591 \times 10^{-4}) \times 10^8 = 6.591 \times 10^4$

c.　$0.0231 \times 10^4 = (2.31 \times 10^{-2}) \times 10^4 = 2.31 \times 10^2$

d.　$0.00000001519 \times 10^{-6} = (1.519 \times 10^{-8}) \times 10^{-6} = 1.519 \times 10^{-14}$

e.　$12.94 \times 10^{-3} = (1.294 \times 10^1) \times 10^{-3} = 1.294 \times 10^{-2}$

f.　1.921×10^{-4} is already in standard scientific notation.

g.　$1 = 1 \times 10^0$

h.　$7,354,211 \times 10^8 = (7.354211 \times 10^6) \times 10^8 = 7.354211 \times 10^{14}$

14.　a.　$1/0.00032 = 3.1 \times 10^3$

b.　$10^3/10^{-3} = 1 \times 10^6$

c.　$10^3/10^3 = 1$ (1×10^0); any number divided by itself is unity.

d.　$1/55,000 = 1.8 \times 10^{-5}$

e. $(10^5)(10^4)(10^{-4})/10^{-2} = 1 \times 10^7$

f. $43.2/(4.32 \times 10^{-5}) = \dfrac{4.32 \times 10^1}{4.32 \times 10^{-5}} = 1.00 \times 10^6$

g. $(4.32 \times 10^{-5})/432 = \dfrac{4.32 \times 10^{-5}}{4.32 \times 10^2} = 1.00 \times 10^{-7}$

h. $1/(10^5)(10^{-6}) = 1/(10^{-1}) = 1 \times 10^1$

16.   grams

18.   a.   mega-

      b.   milli-

      c.   nano-

      d.   mega-

      e.   centi-

      f.   micro-

20.   slightly more than a pound

22.   approximately 1 cup (about half a pint)

24.   kilogram

26.   1.62 m is approximately 5 ft, 4 in. The woman is slightly taller.

28.   a.   centimeter

      b.   meter

      c.   kilometer

30.   d (the other units would give very large numbers for the distance)

32.   Table 2.6 indicates that the diameter of a quarter is 2.5 cm.

$$1 \text{ m} \times \frac{100 \text{ cm}}{1 \text{ m}} \times \frac{1 \text{ quarter}}{2.5 \text{ cm}} = 40 \text{ quarters}$$

34.   uncertainty

36.   The scale of the ruler shown is only marked to the nearest *tenth* of a centimeter; writing 2.850 would imply that the scale was marked to the nearest *hundredth* of a centimeter (and that the zero in the thousandths place had been estimated).

38. a.   probably only two

     b.   infinite (a definition)

     c.   infinite (a definition)

     d.   probably only 1

     e.   three (the race is defined to be exactly 500. miles)

40. final

42. a. 4.23×10^{-1}

     b. 7.12×10^{6}

     c. 4.45×10^{-4}

     d. 2.30×10^{-4}

     e. 9.72×10^{5}

44. a. 3.42×10^{-4}

     b. 1.034×10^{4}

     c. 1.7992×10^{1}

     d. 3.37×10^{5}

46. decimal

48. Most calculators would display 0.66666666. If the 2 and 3 were experimentally determined numbers, this quotient would imply far too many significant figures.

50. none

52. a.   641.0 (the answer can only be given to one decimal place, since 212.7 and 26.7 are only given to one decimal place)

     b.   1.327 (the answer can only be given to three decimal places, since 0.221 is only given to three decimal places)

     c.   77.34 (the answer can only be given to two decimal places, since 26.01 is only given to two decimal places)

     d.   Before performing the calculation, the numbers have to be converted so that they contain the same power of ten.
          $2.01 \times 10^{2} + 3.014 \times 10^{3} = 2.01 \times 10^{2} + 30.14 \times 10^{2} = 32.15 \times 10^{2}$
          This answer should then be converted to *standard* scientific notation, $32.15 \times 10^{2} = 3.215 \times 10^{3} = 3,215$.

54. a.   one (the factor of 2 has only one significant figure)

     b.   four (the sum within the parentheses will contain four significant figures)

     c.   two (based on the factor 4.7×10^{-6} only having two significant figures)

     d.   three (based on the factor 63.9 having only three significant figures)

56. a. $(2.0944 + 0.0003233 + 12.22)/7.001 =$
           $(14.3147233)/7.001 = 2.045$

     b. $(1.42 \times 10^2 + 1.021 \times 10^3)/(3.1 \times 10^{-1}) =$
           $(142 + 1021)/(3.1 \times 10^{-1}) = (1163)/(3.1 \times 10^{-1}) = 3751 = 3.8 \times 10^3$

     c. $(9.762 \times 10^{-3})/(1.43 \times 10^2 + 4.51 \times 10^1) =$
           $(9.762 \times 10^{-3})/(143 + 45.1) = (9.762 \times 10^{-3})/(188.1) = 5.19 \times 10^{-5}$

     d. $(6.1982 \times 10^{-4})^2 = (6.1982 \times 10^{-4})(6.1982 \times 10^{-4}) = 3.8418 \times 10^{-7}$

58. an infinite number (a definition)

60. $\dfrac{1000 \text{ mL}}{1 \text{ L}}$; $\dfrac{1 \text{ L}}{1000 \text{ mL}}$

62. $\dfrac{1 \text{ lb}}{\$0.79}$

64. a. $2.23 \text{ m} \times \dfrac{1.094 \text{ yd}}{1 \text{ m}} = 2.44 \text{ yd}$

     b. $46.2 \text{ yd} \times \dfrac{1 \text{ m}}{1.094 \text{ yd}} = 42.2 \text{ m}$

     c. $292 \text{ cm} \times \dfrac{1 \text{ in.}}{2.54 \text{ cm}} = 115 \text{ in.}$

     d. $881.2 \text{ in.} \times \dfrac{2.54 \text{ cm}}{1 \text{ in.}} = 2238 \text{ cm}$

     e. $1043 \text{ km} \times \dfrac{1 \text{ mi}}{1.6093 \text{ km}} = 648.1 \text{ mi}$

     f. $445.5 \text{ mi} \times \dfrac{1.6093 \text{ km}}{1 \text{ mi}} = 716.9 \text{ km}$

     g. $36.2 \text{ m} \times \dfrac{1 \text{ km}}{1000 \text{ m}} = 0.0362 \text{ km}$

h. $0.501 \text{ km} \times \dfrac{1000 \text{ m}}{1 \text{ km}} \times \dfrac{100 \text{ cm}}{1 \text{ m}} = 5.01 \times 10^4 \text{ cm}$

66.   a. $254.3 \text{ g} \times \dfrac{1 \text{ kg}}{1000 \text{ g}} = 0.2543 \text{ kg}$

     b. $2.75 \text{ kg} \times \dfrac{1000 \text{ g}}{1 \text{ kg}} = 2.75 \times 10^3 \text{ g}$

     c. $2.75 \text{ kg} \times \dfrac{1 \text{ lb}}{0.45359 \text{ kg}} = 6.06 \text{ lb}$

     d. $2.75 \text{ kg} \times \dfrac{1 \text{ lb}}{0.45359 \text{ kg}} \times \dfrac{16 \text{ oz}}{1 \text{ lb}} = 97.0 \text{ oz}$

     e. $534.1 \text{ g} \times \dfrac{1 \text{ lb}}{453.59 \text{ g}} = 1.177 \text{ lb}$

     f. $1.75 \text{ lb} \times \dfrac{453.59 \text{ g}}{1 \text{ lb}} = 794 \text{ g}$

     g. $8.7 \text{ oz} \times \dfrac{1 \text{ lb}}{16 \text{ oz}} \times \dfrac{453.59 \text{ g}}{1 \text{ lb}} = 2.5 \times 10^2 \text{ g}$

     h. $45.9 \text{ g} \times \dfrac{1 \text{ lb}}{453.59 \text{ g}} \times \dfrac{16 \text{ oz}}{1 \text{ lb}} = 1.62 \text{ oz}$

68. $190 \text{ mi} = 1.9 \times 10^2 \text{ mi}$ to two significant figures

     $1.9 \times 10^2 \text{ mi} \times \dfrac{1 \text{ km}}{0.62137 \text{ mi}} = 3.1 \times 10^2 \text{ km}$

     $3.1 \times 10^2 \text{ km} \times \dfrac{1000 \text{ m}}{1 \text{ km}} = 3.1 \times 10^5 \text{ m}$

     $1.9 \times 10^2 \text{ mi} \times \dfrac{5,280 \text{ ft}}{1 \text{ mi}} = 1.0 \times 10^6 \text{ ft}$

70. $1 \times 10^{-10} \text{ m} \times \dfrac{100 \text{ cm}}{1 \text{ m}} = 1 \times 10^{-8} \text{ cm}$

     $1 \times 10^{-8} \text{ cm} \times \dfrac{1 \text{ in.}}{2.54 \text{ cm}} = 4 \times 10^{-9} \text{ in.}$

     $1 \times 10^{-8} \text{ cm} \times \dfrac{1 \text{ m}}{100 \text{ cm}} \times \dfrac{10^9 \text{ nm}}{1 \text{ m}} = 0.1 \text{ nm}$

72.   freezing

74.   273

76.   Fahrenheit (F)

78. $T_C = T_K - 273$

    a.   −210 + 273 = 63 K

    b.   275 K − 273 = 2 °C

    c.   778 K − 273 = 505 °C

    d.   778 + 273 = 1051 K

80. $T_F = 1.80(T_C) + 32$

    a.   1.80(78.1) + 32 = 173 °F

    b.   1.80(40.) + 32 = 104 °F

    c.   1.80(−273) + 32 = −459 °F

    d.   1.80(32) + 32 = 90. °F

82. $T_F = 1.80(T_C) + 32$ $T_C = (T_F - 32)/1.80$

    a.   275 − 273 = 2 °C

    b.   (82 − 32)/1.80 = 28°C

    c.   1.80(−21) + 32 = −5.8°F (−6 °F)

    d.   (−40 − 32)/1.80 = −40 °C (the Celsius and Fahrenheit temperatures
                                      are the same).

84. g/cm^3 (g/mL)

86.   100 $in.^3$

88.   Density is a *characteristic* property, which is always the same for a
     pure substance.

90.   copper

92. $\text{density} = \dfrac{\text{mass}}{\text{volume}}$

    a. $d = \dfrac{122.4 \text{ g}}{5.5 \text{ cm}^3} = 22 \text{ g/cm}^3$

b. $v = 0.57 \text{ m}^3 \times (\dfrac{100 \text{ cm}}{1 \text{ m}})^3 = 5.7 \times 10^5 \text{ cm}^3$

$d = \dfrac{1.9302 \times 10^4 \text{ g}}{5.7 \times 10^5 \text{ cm}^3} = 0.034 \text{ g/cm}^3$

c. $m = 0.0175 \text{ kg} \times \dfrac{1000 \text{ g}}{1 \text{ kg}} = 17.5 \text{ g}$

$d = \dfrac{17.5 \text{ g}}{18.2 \text{ mL}} = 0.962 \text{ g/mL} = 0.962 \text{ g/cm}^3$

d. $v = 0.12 \text{ m}^3 \times (\dfrac{100 \text{ cm}}{1 \text{ m}})^3 = 1.2 \times 10^5 \text{ cm}^3$

$d = \dfrac{2.49 \text{ g}}{1.2 \times 10^5 \text{ cm}^3} = 2.1 \times 10^{-5} \text{ g/cm}^3$

94. $125 \text{ mL} = 125 \text{ cm}^3;$ $125 \text{ mL} \times \dfrac{13.6 \text{ g}}{1 \text{ cm}^3} = 1.70 \times 10^3 \text{ g}$

96. $m = 3.5 \text{ lb} \times \dfrac{453.59 \text{ g}}{1 \text{ lb}} = 1.59 \times 10^3 \text{ g}$

$v = 1.2 \times 10^4 \text{ in.}^3 \times (\dfrac{2.54 \text{ cm}}{1 \text{ in.}})^3 = 1.97 \times 10^5 \text{ cm}^3$

$d = \dfrac{1.59 \times 10^3 \text{ g}}{1.97 \times 10^5 \text{ cm}^3} = 8.1 \times 10^{-3} \text{ g/cm}^3$

The material will float.

98. $5.25 \text{ g} \times \dfrac{1 \text{ cm}^3}{10.5 \text{ g}} = 0.500 \text{ cm}^3 = 0.500 \text{ mL}$

$11.2 \text{ mL} + 0.500 \text{ mL} = 11.7 \text{ mL}$

100.   a. $50.0 \text{ cm}^3 \times \dfrac{19.32 \text{ g}}{1 \text{ cm}^3} = 966 \text{ g}$

b. $50.0 \text{ cm}^3 \times \dfrac{7.87 \text{ g}}{1 \text{ cm}^3} = 394 \text{ g}$

c. $50.0 \text{ cm}^3 \times \dfrac{11.34 \text{ g}}{1 \text{ cm}^3} = 567 \text{ g}$

d. $50.0 \text{ cm}^3 \times \dfrac{2.70 \text{ g}}{1 \text{ cm}^3} = 135 \text{ g}$

102. a. 3.011×10^{23} = 301,100,000,000,000,000,000,000

      b. 5.091×10^{9} = 5,091,000,000

      c. 7.2×10^{2} = 720

      d. 1.234×10^{5} = 123,400

      e. 4.32002×10^{-4} = 0.000432002

      f. 3.001×10^{-2} = 0.03001

      g. 2.9901×10^{-7} = 0.00000029901

      h. 4.2×10^{-1} = 0.42

104. a.   centimeters

      b.   meters

      c.   kilometers

      d.   centimeters

      e.   millimeters

106. a. $36.2 \text{ blim} \times \dfrac{1400 \text{ kryll}}{1 \text{ blim}} = 5.07 \times 10^{4} \text{ kryll}$

      b. $170 \text{ kryll} \times \dfrac{1 \text{ blim}}{1400 \text{ kryll}} = 0.12 \text{ blim}$

      c. $72.5 \text{ kryll}^2 \times \left(\dfrac{1 \text{ blim}}{1400 \text{ kryll}}\right)^2 = 3.70 \times 10^{-5} \text{ blim}^2$

108. $52 \text{ cm} \times \dfrac{1 \text{ in.}}{2.54 \text{ cm}} = 20. \text{ in.}$

110. $1 \text{ lb} \times \dfrac{1 \text{ kg}}{2.2 \text{ lb}} \times \dfrac{\$1}{F5} \times \dfrac{11.5F}{1 \text{ kg}} = \$1$

112. $°X = 1.26°C + 14$

114. $d = \dfrac{36.8 \text{ g}}{10.5 \text{ L}} = 3.50 \text{ g/L} \quad (3.50 \times 10^{-3} \text{ g/cm}^3)$

116. for ethanol, $100. \text{ mL} \times \dfrac{0.785 \text{ g}}{1 \text{ mL}} = 78.5 \text{ g}$

      for benzene, $1000 \text{ mL} \times \dfrac{0.880 \text{ g}}{1 \text{ mL}} = 880. \text{ g}$

total mass, 78.5 + 880. = 959 g

118. a.   negative

b.   negative

c.   positive

d.   zero

e.   negative

120. a.   2; positive

b.   11; negative

c.   3; positive

d.   5; negative

e.   5; positive

f.   0; zero

g.   1; negative

h.   7; negative

122. a.   1; positive

b.   3; negative

c.   0; zero

d.   3; positive

e.   9; negative

124. a.   The decimal point must be moved five places to the left; $2.98 \times 10^{-5} = 0.0000298$.

b.   The decimal point must be moved nine places to the right; $4.358 \times 10^{9} = 4,358,000,000$.

c.   The decimal point must be moved six places to the left; $1.9928 \times 10^{-6} = 0.0000019928$.

d.   The decimal point must be moved 23 places to the right; $6.02 \times 10^{23} = 602,000,000,000,000,000,000,000$.

e.   The decimal point must be moved one place to the left; $1.01 \times 10^{-1} = 0.101$.

f.   The decimal point must be moved three places to the left; $7.87 \times 10^{-3} = 0.00787$.

g.   The decimal point must be moved seven places to the right; $9.87 \times 10^{7} = 98,700,000$.

h.   The decimal point must be moved two places to the right; $3.7899 \times 10^2 = 378.99$.

i.   The decimal point must be moved one place to the left; $1.093 \times 10^{-1} = 0.1093$.

j.   The decimal point must be moved zero places; $2.9004 \times 10^0 = 2.9004$.

k.   The decimal point must be moved four places to the left; $3.9 \times 10^{-4} = 0.00039$.

l.   The decimal point must be moved eight places to the left; $1.904 \times 10^{-8} = 0.00000001904$.

126. a. $1/10^2 = 1 \times 10^{-2}$

b. $1/10^{-2} = 1 \times 10^2$

c. $55/10^3 = \dfrac{5.5 \times 10^1}{1 \times 10^3} = 5.5 \times 10^{-2}$

d. $(3.1 \times 10^6)/10^{-3} = \dfrac{3.1 \times 10^6}{1 \times 10^{-3}} = 3.1 \times 10^9$

e. $(10^6)^{1/2} = 1 \times 10^3$

f. $(10^6)(10^4)/(10^2) = \dfrac{(1 \times 10^6)(1 \times 10^4)}{(1 \times 10^2)} = 1 \times 10^8$

g. $1/0.0034 = \dfrac{1}{3.4 \times 10^{-3}} = 2.9 \times 10^2$

h. $3.453/10^{-4} = \dfrac{3.453}{1 \times 10^{-4}} = 3.453 \times 10^4$

128. Kelvin, K

130. centimeter

132. 0.105 m

134. 1 kg (100 g = 0.1 kg)

136. 10 cm (1 cm = 10 mm)

138. 2.8 (the hundredths place is estimated)

140. a.   0.000426

      b. 4.02×10^{-5}

      c. 5.99×10^{6}

      d.   400.

      e.   0.00600

142. a.   2149.6 (the answer can only be given to the first decimal place, because 149.2 is only known to the first decimal place)

      b. 5.37×10^{3} (the answer can only be given to two decimal places because 4.34 is only known to two decimal places; moreover, since the power of ten is the same for each number, the calculation can be performed directly)

      c.   Before performing the calculation, the numbers have to be converted so that they contain the same power of ten.

            $4.03 \times 10^{-2} - 2.044 \times 10^{-3} =$

            $4.03 \times 10^{-2} - 0.2044 \times 10^{-2} =$

            3.83×10^{-2} (the answer can only be given the second decimal place because 4.03×10^{-2} is only known to the second decimal place)

      d.   Before performing the calculation, the numbers have to be converted so that they contain the same power of ten.

            $2.094 \times 10^{5} - 1.073 \times 10^{6} =$

            $2.094 \times 10^{5} - 10.73 \times 10^{5} =$

            -8.64×10^{5}

144. a. $(2.9932 \times 10^{4})(2.4443 \times 10^{2} + 1.0032 \times 10^{1}) =$

            $(2.9932 \times 10^{4})(24.443 \times 10^{1} + 1.0032 \times 10^{1}) =$

            $(2.9932 \times 10^{4})(25.446 \times 10^{1}) = 7.6166 \times 10^{6}$

      b. $(2.34 \times 10^{2} + 2.443 \times 10^{-1})/(0.0323) =$

            $(2.34 \times 10^{2} + 0.002443 \times 10^{2})/(0.0323) =$

            $(2.34 \times 10^{2})/(0.0323) = 7.24 \times 10^{3}$

      c. $(4.38 \times 10^{-3})^{2} = 1.92 \times 10^{-5}$

      d. $(5.9938 \times 10^{-6})^{1/2} = 2.4482 \times 10^{-3}$

146.   1 year/12 months; 12 months/1 year

148.   a. $908 \text{ oz} \times \dfrac{1 \text{ lb}}{16 \text{ oz}} \times \dfrac{1 \text{ kg}}{2.2046 \text{ lb}} = 25.7 \text{ kg}$

   b. $12.8 \text{ L} \times \dfrac{1 \text{ qt}}{0.94633 \text{ L}} \times \dfrac{1 \text{ gal}}{4 \text{ qt}} = 3.38 \text{ gal}$

   c. $125 \text{ mL} \times \dfrac{1 \text{ L}}{1000 \text{ mL}} \times \dfrac{1 \text{ qt}}{0.94633 \text{ L}} = 0.132 \text{ qt}$

   d. $2.89 \text{ gal} \times \dfrac{4 \text{ qt}}{1 \text{ gal}} \times \dfrac{1 \text{ L}}{1.0567 \text{ qt}} \times \dfrac{1000 \text{ mL}}{1 \text{ L}} = 1.09 \times 10^4 \text{ mL}$

   e. $4.48 \text{ lb} \times \dfrac{453.59 \text{ g}}{1 \text{ lb}} = 2.03 \times 10^3 \text{ g}$

   f. $550 \text{ mL} \times \dfrac{1 \text{ L}}{1000 \text{ mL}} \times \dfrac{1.0567 \text{ qt}}{1 \text{ L}} = 0.58 \text{ qt}$

150.   Assuming exactly 6 gross, 864 pencils

152.   a.   Celsius temperature = (175 − 32)/1.80 = 79.4 °C

   Kelvin temperature = 79.4 + 273 = 352 K

   b.   255 − 273 = −18 °C

   c.   (−45 − 32)/1.80 = −43 °C

   d.   1.80(125) + 32 = 257 °F

154. $85.5 \text{ mL} \times \dfrac{0.915 \text{ g}}{1 \text{ mL}} = 78.2 \text{ g}$

156. $m = 155 \text{ lb} \times \dfrac{453.59 \text{ g}}{1 \text{ lb}} = 7.031 \times 10^4 \text{ g}$

   $v = 4.2 \text{ ft}^3 \times \left(\dfrac{12 \text{ in.}}{1 \text{ ft}}\right)^3 \times \left(\dfrac{2.54 \text{ cm}}{1 \text{ in.}}\right)^3 = 1.189 \times 10^5 \text{ cm}^3$

   $d = \dfrac{7.031 \times 10^4 \text{ g}}{1.189 \times 10^5 \text{ cm}^3} = 0.59 \text{ g/cm}^3$

158.   a. 23 °F

   b. 32 °F

   c. −321 °F

   d. −459 °F

   e. 187 °F

   f. −459 °F

Chapter 3   Matter and Energy

2.   solid, liquid, gas (vapor)

4.   Liquids

6.   gaseous

8.   The stronger the interparticle forces, the more rigid is the sample overall.

10. Gases are easily compressed into smaller volumes, whereas solids and liquids are not. Because a gaseous sample consists mostly of empty space, it is this empty space that is compressed when pressure is applied to a gas.

12. chemical

14. Nickel(II) sulfate *reacts* with barium chloride.

16. The most common physical changes are changes in state: solid to liquid, liquid to gaseous, solid to gaseous (and their opposites).

18. a.   chemical;   the scorch represents the oxidation of the material

   b.   physical;   the gas in the tires decreases in volume with temperature

   c.   chemical;   tarnish on silver is caused by reaction of the silver with sulfur or oxygen

   d.   chemical;   the ethyl alcohol in the wine is oxidized to acetic acid

   e.   chemical;   the oven cleaner contains sodium hydroxide that converts greases in the oven into soaps

   f.   chemical;   an ordinary flashlight battery is constructed with a zinc casing, which serves as one of the electrodes. As the battery discharges, the zinc is oxidized.

   g.   chemical;   the acids attack the calcium phosphate matrix of the teeth

   h.   chemical;   the charring represents the breakdown of the sugar

   i.   chemical;   iron in the blood catalyzes the decomposition of the hydrogen peroxide into oxygen gas (and water)

   j.   physical;   this is only a change in state; carbon dioxide, only in the gaseous state, is still present after the sublimation

   k.   chemical;   chlorine is an oxidizing agent and can change the chemical nature of the dyes in the fabrics

20.   element

22.   compounds

24.   Typically, the properties of a compound and the elements that constitute it are very different. Consider the properties of liquid *water* and the hydrogen and oxygen gases from which the water was prepared. Consider the properties of *sodium chloride* (table salt) and the sodium metal and chlorine gas from which it might have been prepared.

26.   Mixtures have variable compositions. A given pure substance has a fixed composition.

28.   solutions: window cleaner, shampoo, rubbing alcohol
       mixtures: salad dressing, jelly beans, the change in my pocket

30.   a.   mixture   c.   mixture

       b.   mixture   d.   pure substance

32.   a.   homogeneous   c.   heterogeneous

       b.   heterogeneous   d.   homogeneous

       e.   The paper itself is basically homogeneous in appearance.

34.   Consider a mixture of salt (sodium chloride) and sand. Salt is soluble in water, sand is not. The mixture is added to water and stirred to dissolve the salt and is then filtered. The salt solution passes through the filter; the sand remains on the filter. The water can then be evaporated from the salt.

36.   The solution is heated to vaporize (boil) the water. The water vapor is then cooled so that it condenses back to the liquid state, and the liquid is collected. After all the water is vaporized from the original sample, pure sodium chloride will remain. The process consists of physical changes.

38.   1 cal = 4.184 Joule, so the calorie is about 4 times larger.

40.   As the steam is cooled from 150 °C to 100 °C, the molecules of vapor gradually slow down as they lose kinetic energy. At 100 °C, the steam condenses into liquid water, and the temperature remains at 100 °C until all the steam has condensed. As the liquid water cools, the molecules in the liquid move more and more slowly as they lose kinetic energy. At 0°C, the liquid water freezes.

42.   temperature

44. $526 \text{ J} \times \dfrac{55 \text{ °C}}{17 \text{ °C}} = 1.7 \times 10^3 \text{ J}$

46.   Because 1.000 cal = 4.184 J, then 1.000 kcal = 4.184 kJ

　　　a.　462.4 kJ × $\dfrac{1\ \text{kcal}}{4.184\ \text{kJ}}$ = 110.5 kcal

　　　b.　18.28 kJ × $\dfrac{1\ \text{kcal}}{4.184\ \text{kJ}}$ = 4.369 kcal

　　　c.　1.014 kJ × $\dfrac{1\ \text{kcal}}{4.184\ \text{kJ}}$ = 0.2424 kcal

　　　d.　190.5 kJ × $\dfrac{1\ \text{kcal}}{4.184\ \text{kJ}}$ = 45.53 kcal

48.　a.　12.30 kcal × $\dfrac{1000\ \text{cal}}{1\ \text{kcal}}$ = 12,300 cal (1.230 × 10^4 cal)

　　　b.　290.4 kcal × $\dfrac{1000\ \text{cal}}{1\ \text{kcal}}$ = 290,400 cal (2.904 × 10^5 cal)

　　　c.　940,000 kcal × $\dfrac{1000\ \text{cal}}{1\ \text{kcal}}$ = 940,000,000 cal (9.4 × 10^8 cal)

　　　d.　4201 kcal × $\dfrac{1000\ \text{cal}}{1\ \text{kcal}}$ = 4,201,000 cal (4.201 × 10^6 cal)

50.　a.　76.52 cal × $\dfrac{1\ \text{kcal}}{1000\ \text{cal}}$ × $\dfrac{4.184\ \text{kJ}}{1\ \text{kcal}}$ = 0.3202 kJ

　　　b.　7.824 kJ × $\dfrac{1\ \text{kcal}}{4.184\ \text{kJ}}$ = 1.870 kcal

　　　c.　489.4 J × $\dfrac{1\ \text{cal}}{4.184\ \text{J}}$ = 117.0 cal

　　　d.　1.598 × 10^4 J × $\dfrac{1\ \text{kJ}}{1000\ \text{J}}$ × $\dfrac{1\ \text{kcal}}{4.184\ \text{kJ}}$ = 3.819 kcal

52.   Heat = mass × specific heat capacity × temperature change

　　　specific heat capacity = Heat/(mass × temperature change)

　　　72.4 kJ = 72,400 J

　　　specific heat capacity = $\dfrac{72,400\ \text{J}}{(952\ \text{g})(10.7\ °\text{C})}$ = 7.11 J/g °C

54.   Table 3.2 gives the specific heat of iron as 0.45 J/g °C.

　　　Temperature increase = 50.0 − 25.0 = 25.0 °C

　　　562 J = (mass of iron) × 0.45 J/g °C × 25.0 °C

　　　mass of iron = 50. g

56.   Table 3.2 gives the specific heat capacity of iron as 0.45 J/g °C.

Heat = mass × specific heat capacity × temperature change

Temperature change = 75.5 - 40.1 = 35.4 °C

Heat = 852.2 g × 0.45 J/g °C × 35.4 °C = 1.4×10^4 J

58. $0.13 \dfrac{J}{g\ °C} \times \dfrac{1\ cal}{4.184\ J} = 0.031 \dfrac{cal}{g\ °C}$

60.   Specific heat capacities are given in Table 3.2

for gold,   25.0 g × 0.13 J/g °C × 20. °C = 65 J

for mercury,   25.0 g × 0.14 J/g °C × 20. °C = 70. J

for carbon,   25.0 g × 0.71 J/g °C × 20. °C = 360 J

62.   1251 J = 35.2 g × (specific heat capacity) × 25.0 °C

specific heat capacity = 1.42 J/g °C

64.   Since *X* is a pure substance, the fact that two different solids form when electrical current is passed indicates that *X* must be a compound.

66.   Because vaporized water is still the *same substance* as solid water, no chemical reaction has occurred. Sublimation is a physical change.

68.   2.5 kg of water = 2,500 g

Temperature change = 55.0 - 18.5 = 36.5 °C

2500 g × 4.184 J/g °C × 36.5 °C = 3.8×10^5 J

70.   No calculation is necessary: aluminum will lose more heat because it has the higher specific heat capacity.

72.   For any substance, $Q = m \times s \times \Delta T$. The quantity of heat gained by the water in this experiment must equal the total heat lost by the metals (i.e., the *sum* of the amount of heat lost by the iron and the amount of heat lost by the aluminum).

$(m \times s \times \Delta T)_{water} = (m \times s \times \Delta T)_{iron} + (m \times s \times \Delta T)_{aluminum}$

If T_f represents the final temperature reached by this system, then

$[m \times s \times (T_f - 22.5\ °C)]_{water}$

$= [m \times s \times (100\ °C - T_f)]_{iron} + [m \times s \times (100\ °C - T_f)]_{aluminum}$

$[97.3 \text{ g} \times 4.184 \text{ J/g } °C \times (T_f - 22.5 °C)]$

$= [10.00 \text{ g} \times 0.45 \text{ J/g } °C \times (100 °C - T_f)]$

$+ [5.00 \text{ g} \times 0.89 \text{ J/g } °C \times (100 °C - T_f)]$

$407.1(T_f - 22.5) = 4.50(100 - T_f) + 4.45(100 - T_f) = 8.95(100 - T_f)$

$407.1 T_f - 9160. = 895 - 8.95 T_f$

$416.1 T_f = 10,055$, which gives $T_f = 24.2 °C$

74. $(m \times s \times \Delta T)_{\text{water}} = (m \times s \times \Delta T)_{\text{iron}}$

Let T_f represent the final temperature reached by the system.

$[75 \text{ g} \times 4.184 \text{ J/g } °C \times (T_f - 20.)] = [25.0 \text{ g} \times 0.45 \text{ J/g } °C \times (85 - T_f)]$

$314(T_f - 20.) = 11.4(85 - T_f)$

$314 T_f - 6280 = 961 - 11.3 T_f$

$325 T_f = 7241$

$T_f = 22.3 °C = 22 °C$

76. far apart

78. chemical

80. chemical

82. electrolysis

84. a.   heterogeneous

b.   heterogeneous

c.   heterogeneous (unless you work hard to get all the lumps out!)

d.   although strictly heterogenous, it may appear homogeneous

e.   heterogeneous

86. 9.0 J (It requires twice as much heat to warm twice as large a sample over the same temperature interval.)

88. a. $44.21 \text{ cal} \times \dfrac{4.184 \text{ J}}{1 \text{ cal}} = 185.0 \text{ J}$

b. $162.4 \text{ cal} \times \dfrac{4.184 \text{ J}}{1 \text{ cal}} = 679.5 \text{ J}$

c. $3.721 \times 10^3 \text{ cal} \times \dfrac{4.184 \text{ J}}{1 \text{ cal}} = 1.557 \times 10^4 \text{ J}$

d. $146.2 \text{ kcal} \times \dfrac{1000 \text{ cal}}{1 \text{ kcal}} \times \dfrac{4.184 \text{ J}}{1 \text{ cal}} = 6.117 \times 10^5 \text{ J}$

90.   a. $5.442 \times 10^4 \text{ J} \times \dfrac{1 \text{ kJ}}{1000 \text{ J}} = 54.42 \text{ kJ}$

b. $5.442 \times 10^4 \text{ J} \times \dfrac{1 \text{ cal}}{4.184 \text{ J}} = 1.301 \times 10^4 \text{ cal}$

c. $352.6 \text{ kcal} \times \dfrac{4.184 \text{ kJ}}{1 \text{ kcal}} = 1475 \text{ kJ}$

d. $17.24 \text{ kJ} \times \dfrac{1 \text{ kcal}}{4.184 \text{ kJ}} = 4.120 \text{ kcal}$

92.   The specific heat capacity of water is 4.184 J/g °C.
Temperature increase = 39 − 25 = 14 °C.
75 g × 4.184 J/g °C × 14 °C = 4400 J (to 2 significant figures)

94.   For any substance, $Q = m \times s \times \Delta T$. The basic calculation for each of the substances is the same (specific heat capacities are found in Table 3.2)

Heat required = 150. g × (specific heat capacity) × 11.2 °C

Substance	Specific Heat Capacity	Heat Required
water (*l*)	4.184 J/g °C	7.03×10^3 J
water (*s*)	2.03 J/g °C	3.41×10^3 J
water (*g*)	2.0 J/g °C	3.4×10^3 J
aluminum	0.89 J/g °C	1.5×10^3 J
iron	0.45 J/g °C	7.6×10^2 J
mercury	0.14 J/g °C	2.4×10^2 J
carbon	0.71 J/g °C	1.2×10^3 J
silver	0.24 J/g °C	4.0×10^2 J
gold	0.13 J/g °C	2.2×10^2 J

96.   Heat = mass × specific heat capacity × temperature change

The temperature change is 112.1 − 25.0 = 87.1 °C.

1.351 kJ = 1351 J

$$s = \dfrac{1351 \text{ J}}{125 \text{ g} \times 87.1 \text{ °C}} = 0.124 \text{ J/g °C}$$

Cumulative Review: Chapters 1, 2, and 3

2.   By now, after having covered three chapters in this book, it is hoped
      that you have adopted an "active" approach to your study of chemistry.
      You may have discovered (perhaps through a disappointing grade on a quiz
      (though we hope not), that you really have to get involved with
      chemistry. You can't just sit and take notes, or just look over the
      solved examples in the textbook. You have to learn to solve problems.
      You have to learn how to interpret problems, and how to reduce them to
      the simple mathematical relationships you have studied. Whereas in some
      courses you might get by on just giving back on exams the facts or ideas
      presented in class, in chemistry you have to be able to extend and
      synthesize what has been discussed and to apply the material to new
      situations. Don't get discouraged if this is difficult at first: it's
      difficult for everyone at first.

4.   It is difficult sometimes for students (especially beginning students)
      to understand why certain subjects are required for a given college
      major. The faculty of your major department, however, have collectively
      many years of experience in the subject in which you have chosen to
      specialize. They really do know what courses will be helpful to you in
      the future. They may have had trouble with the same courses that now
      give you trouble, but they realize that all the work will be worth it in
      the end. Some courses you take, particularly in your major field itself,
      have obvious and immediate utility. Other courses, often times chemistry
      included, are provided to give you a general background knowledge, which
      may prove useful in understanding your own major or other subjects
      related to your major. In perhaps a burst of bravado, chemistry has been
      called "the central science" by one team of textbook authors. This
      moniker is very true however: in order to understand biology, physics,
      nutrition, farming, home economics, or whatever (it helps to have a
      general background in chemistry).

6.   Whenever a scientific measurement is made, we always employ the
      instrument or measuring device we are using to the limits of its
      precision. On a practical basis, this usually means that we *estimate* our
      reading of the last significant figure of the measurement. An example of
      the uncertainty in the last significant figure is given for measuring
      the length of a pin in the text in Figure 2.5. Scientists appreciate the
      limits of experimental techniques and instruments, and always assume
      that the last digit in a number representing a measurement has been
      estimated. Because the last significant figure in every measurement is
      assumed to be estimated, it is never possible to exclude uncertainty
      from measurements. The best we can do is to try to improve our
      techniques and instruments so that we get more significant figures for
      our measurements.

8.   Dimensional analysis is a method of problem solving that pays particular
      attention to the units of measurements and uses these units as if they
      were algebraic symbols that multiply, divide, and cancel. Consider the
      following example. A dozen eggs costs $1.25. Suppose we want to know how
      much one egg costs, and also how much three dozens of eggs will cost. To
      solve these problems, we need to make use of two equivalence statements:

1 dozen eggs = 12 eggs

1 dozen eggs = $1.25

The first of these equivalence statements is obvious: everyone knows that 12 eggs is "equivalent" to one dozen. The second statement also expresses an equivalence: if you give the grocer $1.25, he or she will give you a dozen eggs. From these equivalence statements, we can construct the conversion factors we need to answer the two questions. For the first question (what does one egg cost) we can set up the calculation as follows

$$\frac{\$1.25}{12 \text{ eggs}} = \$0.104 = \$0.10$$

as the cost of one egg. Similarly, for the second question (the cost of 3 dozens eggs), we can set up the conversion as follows

$$3 \text{ dozens} \times \frac{\$1.25}{1 \text{ dozen}} = \$3.75$$

as the cost of three dozens eggs. See Section 2.6 of the text for how we construct conversion factors from equivalence statements.

10.   Defining what scientists mean by "matter" often seems circular to students. Scientists say that matter is something that "has mass and occupies space", without ever really explaining what it means to "have mass" or to "occupy space"! The concept of matter is so basic and fundamental, that it becomes difficult to give a good textbook definition other than to say that matter is the "stuff" of which everything is made. Matter can be classified and subdivided in many ways, depending on what we are trying to demonstrate.

On the most fundamental basis, all matter is composed of tiny particles (such as protons, electrons, neutrons, and the other subatomic particles). On one higher level, these tiny particles are combined in a systematic manner into units called atoms. Atoms, in turn, may be combined to constitute molecules. And finally, large groups of molecules may be placed together to form a bulk sample of substance that we can see.

Matter can also be classified as to the physical state a particular substance happens to take. Some substances are solids, some are liquids, and some are gases. Matter can also be classified as to whether it is a pure substance (one type of molecule) or a mixture (more than one type of molecule), and furthermore whether a mixture is homogeneous or heterogeneous.

12.   Chemists tend to give a functional definition of what they mean by an "element": an element is a fundamental substance that cannot be broken down into any simpler substances by chemical methods. Compounds, on the other hand, can be broken down into simpler substances (the elements of which the compound is composed). For example, sulfur and oxygen are both elements (sulfur occurs as S_8 molecules and oxygen as O_2 molecules).

When sulfur and oxygen are placed together and heated, the compound sulfur dioxide (SO_2) forms. When we analyze the sulfur dioxide produced, we notice that each and every molecule consists of one sulfur atom and two oxygen atoms, and on a mass basis, consists of 50% each of sulfur and oxygen. We describe this by saying that sulfur dioxide has a constant composition. The fact that a given compound has constant composition is usually expressed in terms of the mass percentages of the elements present in the compound. The reason the mass percentages are constant is because of a constant number of atoms of each type present in the compound's molecules. If a scientist anywhere in the universe analyzed sulfur dioxide, he or she would find the same composition: if a scientist finds something that does not have the same composition, then the substance cannot be sulfur dioxide.

14.   Scientists define energy as "the capacity to do work". As with trying to define "matter" earlier in this chapter, energy is such a fundamental concept that it is hard to define (what is "work"?). Although the *SI* unit of energy is the *joule*, until relatively recently, energies were more commonly given in terms of the *calorie*: one calorie is defined to be the amount of heat required to raise the temperature of one gram of water by one Celsius degree. The calorie is a "working" definition, and we can more easily appreciate this amount of energy. In terms of the *SI* unit, 1 calorie = 4.184 joule, so it takes 4.184 J to raise the temperature of one gram of water by one Celsius degree.

The specific heat capacity of a substance, in general, is the amount of energy required to raise the temperature of one gram of a substance by one Celsius degree. Therefore, the specific heat capacity of water must be 1.000 cal/g°C or 4.184 J/g°C. To see how specific heat capacities may be used to calculate the energy change for a process, consider this example: How much energy is required to warm 25.0 g of water from 15.1°C to 35.2°C? The specific heat capacity of water is 1.000 cal/g°C; this is the quantity of energy required to raise the temperature of only *one* gram of water by only *one* Celsius degree. In this example, we are raising the temperature of 25.0 g of water, and we are raising the temperature by (35.2 - 15.1) = 20.1°C. So, using the specific heat capacity as a conversion factor, we can say

$$\text{energy required} = \left(\frac{1.00 \text{ cal}}{\text{g } °C} \right) \times (25.0 \text{ g}) \times (20.1°C) = 503 \text{ cal}$$

Notice how the units of g and °C cancel, leaving the answer in energy units only. This sort of calculation of energy change can be done for any substance, using the substance's own specific heat capacity (see Table 3.2).

16.   a. $2.4 \text{ L} \times \dfrac{1.0567 \text{ qt}}{1 \text{ L}} = 2.5 \text{ qt}$

b. $785.2 \text{ g} \times \dfrac{1 \text{ lb}}{453.59 \text{ g}} = 1.731 \text{ lb}$

c. $785.2 \text{ g} \times \dfrac{1 \text{ kg}}{1000 \text{ g}} = 0.7852 \text{ kg}$

d. $7521 \text{ m} \times \dfrac{1 \text{ km}}{1000 \text{ m}} \times \dfrac{1 \text{ mi}}{1.6093 \text{ km}} = 4.673 \text{ mi}$

e. $25.2 \text{ cm} \times \dfrac{1 \text{ m}}{100 \text{ cm}} = 0.252 \text{ m}$

f. $451 \text{ mL} \times \dfrac{1 \text{ L}}{1000 \text{ mL}} = 0.451 \text{ L}$

g. $6.45 \text{ m} \times \dfrac{1.0936 \text{ yd}}{1 \text{ m}} \times \dfrac{3 \text{ ft}}{1 \text{ yd}} = 21.2 \text{ ft}$

h. $125.5 \text{ cm} \times \dfrac{1 \text{ in.}}{2.54 \text{ cm}} = 49.41 \text{ in.}$

i. $6.29 \text{ in.} \times \dfrac{1 \text{ ft}}{12 \text{ in.}} = 0.524 \text{ ft}$

j. $4.62 \text{ oz} \times \dfrac{1 \text{ lb}}{16 \text{ oz}} = 0.289 \text{ lb}$

18. $T_F = 1.80(T_C) + 32 \qquad T_C = (T_F - 32)/1.80 \qquad T_K = T_C + 273$

a. $1.80(-40 \text{ °C}) + 32 = -40 \text{ °F}$

b. $577 \text{ K} - 273 = 304 \text{ °C}$

c. $(225 \text{ °F} - 32)/1.80 = 107 \text{ °C}$

d. $-158 \text{ °C} + 273 = 115 \text{ K}$

e. $(25.9 \text{ °F} - 32)/1.80 = -3.4 \text{ °C}$

f. $1.80(25.9 \text{ °C}) + 32 = 78.6 \text{ °F}$

20.   a. $425 \text{ J} \times \dfrac{1 \text{ kJ}}{1000 \text{ J}} = 0.425 \text{ kJ}$

b. $425 \text{ J} \times \dfrac{1 \text{ cal}}{4.184 \text{ J}} = 102 \text{ cal}$

c.   425 J = 102 cal (part b) = 0.102 kcal

d. $78.5 \text{ kcal} \times \dfrac{4.184 \text{ kJ}}{1 \text{ kcal}} \times \dfrac{1000 \text{ J}}{1 \text{ kJ}} = 3.28 \times 10^5 \text{ J}$

e. $78.5 \text{ kcal} \times \dfrac{4.184 \text{ kJ}}{1 \text{ kcal}} = 328 \text{ kJ}$

f. $78.5 \text{ kcal} \times \dfrac{1000 \text{ cal}}{1 \text{ kcal}} = 7.85 \times 10^4 \text{ cal}$

Chapter 4   Chemical Foundations: Elements, Atoms, and Ions

2.   Robert Boyle

4.   There are 112 elements presently known; of these 88 occur naturally and 24 are manmade. Table 4.1 lists the most common elements on the Earth.

6.   The four most abundant elements in living creatures are, respectively, oxygen, carbon, hydrogen, and nitrogen (see Table 4.2). In the nonliving world, the four most abundant elements are, respectively, oxygen, silicon, aluminum, and iron (see Table 4.1).

8.   Sb (antimony)

      Cu (copper)

      Au (gold)

      Pb (lead)

      Hg (mercury)

      K (potassium)

      Ag (silver)

      Na (sodium)

      Sn (tin)

      W (tungsten)

      Fe (iron)

10.   a.   4   f.   7

      b.   6   g.   12

      c.   9   h.   2

      d.   1   i.   14

      e.   5   j.   13

12. | Symbol | Name |
      |--------|------|
      | Ru | ruthenium |
      | Tl | thallium |
      | No | nobelium |
      | Th | thorium |
      | Pu | plutonium |
      | Os | osmium |

14. a.   bromine

　　 b.   boron

　　 c.   beryllium

　　 d.   bismuth

　　 e.   barium

16. A given compound is always made up of the same number and type of atoms, and so the composition of the compound on a mass percentage basis will always be the same, no matter what the source of the compound is. For example, water molecules from the Atlantic Ocean and the Pacific Ocean all contain two hydrogen atoms bonded to one oxygen atom. If a molecule does *not* contain two hydrogen atoms bonded to one oxygen atom, then the molecule is not water.

18. According to Dalton, all atoms of the same element are *identical*; in particular, every atom of a given element has the same *mass* as every other atom of that element. If a given compound always contains the *same relative numbers* of atoms of each kind, and those atoms always have the *same masses*, then it follows that the compound made from those elements would always contain the same relative masses of its elements.

20. a. C_6H_6　　　　　　　　d. $FeBr_3$

　　 b. N_2O_4　　　　　　　　e. $NaNO_3$

　　 c. $CaCl_2$　　　　　　　　f. Ca_3N_2

22. a.   False; Rutherford's bombardment experiments with metal foil suggested that the alpha particles were being deflected by coming near a *dense, positively charged* atomic nucleus.

　　 b.   False; The proton and the electron have opposite charges, but the mass of the electron is *much smaller* than the mass of the proton.

　　 c.   True

24. The protons and neutrons are found in the nucleus. The protons are positively charged; the neutrons have no electrical charge. Protons and neutrons each have approximately the same mass.

26. neutron; electron

28. Because they are located in the exterior regions of the atom, it is the electrons of an atom that most interact with other atoms and are therefore most responsible for the atom's chemical behavior.

30. False; the mass number represents the total number of protons and neutrons in the nucleus.

32. mass

34.   Atoms of the same element (i.e., atoms with the same number of protons in the nucleus) may have different numbers of neutrons, and so will have different masses.

36.   a.   32   e.   38

       b.   30   f.   27

       c.   24   g.   4

       d.   74   h.   3

38.   a. $^{58}_{27}\text{Co}$ d. $^{132}_{53}\text{I}$

       b. $^{10}_{5}\text{B}$ e. $^{19}_{9}\text{F}$

       c. $^{23}_{12}\text{Mg}$ f. $^{65}_{29}\text{Cu}$

40.   a.   92 protons, 143 neutrons, 92 electrons

       b.   6 protons, 7 neutrons, 6 electrons

       c.   26 protons, 31 neutrons, 26 electrons

       d.   82 protons, 126 neutrons, 82 electrons

       e.   37 protons, 49 neutrons, 37 electrons

       f.   20 protons, 21 neutrons, 20 electrons

42.

Element	Neutrons	Atomic Number	Mass Number	Symbol
nitrogen	6	7	13	$^{13}_{7}\text{N}$
nitrogen	7	7	14	$^{14}_{7}\text{N}$
lead	124	82	206	$^{206}_{82}\text{Pb}$
iron	31	26	57	$^{57}_{26}\text{Fe}$
krypton	48	36	84	$^{84}_{36}\text{Kr}$

44.   Elements with similar chemical properties are aligned *vertically* in families known as *groups*.

46.   Metallic elements are found towards the *left* and *bottom* of the periodic table; there are far more metallic elements than there are nonmetals.

48.   The gaseous nonmetallic elements are hydrogen, nitrogen, oxygen, fluorine, chlorine, plus all the group 8 elements (noble gases). There are no gaseous metallic elements under room conditions.

50.   The metalloids are the elements found on either side of the "stairstep" region that is marked on most periodic tables. The metalloid elements show some properties of both metals and nonmetals.

52.   a.   Group 1A, alkali metals

       b.   Group 3A

       c.   Group 8A, noble gases

       d.   Group 4A

       e.   Group 6A (sometimes called the "chalcogens")

       f.   Group 8A, noble gases

       g.   Group 6A (sometimes called the "chalcogens")

54.

Name	*Symbol*	*Atomic Number*	*Group Number*	*Metal/Nonmetal*
lithium	Li	3	1A	metal
arsenic	As	33	5A	metalloid
radon	Rn	86	8A	nonmetal
radium	Ra	88	2A	metal
germanium	Ge	32	4A	metalloid

56.   Most of the elements are too reactive to be found in the uncombined form in nature and are found only in compounds.

58.   These elements are found *uncombined* in nature and do not readily react with other elements. For many years it was thought that these elements formed no compounds at all, although this has now been shown to be untrue.

60.   diatomic gases: H_2, N_2, O_2, Cl_2, and F_2

       monatomic gases: He, Ne, Kr, Xe, Rn, and Ar

62.   chlorine

64.   diamond

66.   Positive ions are produced when an atom *loses* one or more electrons. Negative ions are produced when an atom *gains* one or more electrons. Formation of a simple ion does *not* involve any change in the nucleus.

68.   positively

70.   -*ide*

72.   nonmetallic

74.   a. Fe_2S_3 e. Rb_2S

       b.   FeS   f.   AlP

       c. Al_2O_3 g.   MnS

       d. AlI_3 h. SnS_2

76.   a. $2e^-$ d. $4e^-$

       b. $3e^-$ e. $1e^-$

       c. $2e^-$ f. $2e^-$

78.   a. Tl^{3+} d. As^{3-}

       b. Se^{2-} e. Fr^+

       c. Ba^{2+} f. Cs^+

80.   Sodium chloride is an *ionic* compound, consisting of Na^+ and Cl^- *ions*.
       When NaCl is dissolved in water, these ions are *set free*, and can move
       independently to conduct the electrical current. Sugar crystals, al-
       though they may visually *appear* similar contain *no* ions. When sugar is
       dissolved in water, it dissolves as uncharged *molecules*. There are no
       electrically charged species present in a sugar solution to carry the
       electrical current.

82.   The total number of positive charges must equal the total number of nega-
       tive charges so that there will be *no net charge* on the crystals of an
       ionic compound. A macroscopic sample of compound must ordinarily not
       have any net charge.

84.   a.   One 2+ ion is exactly balanced by one 2- ion: CoO.

       b.   The smallest common multiple of 3 and 2 is 6; two 3+ ions are
             balanced by three 2- ions: Co_2S_3.

       c.   Three 1- ions are needed to balance one 3+ ion: $AlCl_3$.

       d.   The smallest common multiple of 3 and 2 is 6; three 2+ ions are
             balanced by two 3- ions: Ba_3N_2.

       e.   One 4- ion would be balanced by two 2+ ions: Ca_2C.

       f.   One 3- ion would be balanced by three 1+ ions; K_3N.

86. a.   7; halogens

     b.   8; noble gases

     c.   2; alkaline earth elements

     d.   2; alkaline earth elements

     e.   4

     f.   6; (the members of group 6 are sometimes called the chalcogens)

     g.   8; noble gases

     h.   1; alkali metals

88.

Group 3	*Element*	*Symbol*	*Atomic Number*
	boron	B	5
	aluminum	Al	13
	gallium	Ga	31
	indium	In	49
Group 5	nitrogen	N	7
	phosphorus	P	15
	arsenic	As	33
	antimony	Sb	51
Group 8	helium	He	2
	neon	Ne	10
	argon	Ar	18
	krypton	Kr	36

90. Most of the mass of an atom is concentrated in the nucleus: the *protons* and *neutrons* that constitute the nucleus have similar masses, and these particles are nearly two thousand times heavier than electrons. The chemical properties of an atom depend on the number and location of the *electrons* it possesses. Electrons are found in the outer regions of the atom and are the particles most likely to be involved in interactions between atoms.

92. $C_6H_{12}O_6$

94. a.   29 protons; 34 neutrons; 29 electrons

     b.   35 protons; 45 neutrons; 35 electrons

     c.   12 protons; 12 neutrons; 12 electrons

96. The chief use of gold in ancient times was as *ornamentation*, whether in statuary or in jewelry. Gold possesses an especially beautiful luster, and because it is relatively soft and malleable, it could be worked finely by

artisans. Among the metals, gold is particularly inert to attack by most substances in the environment.

98.   a.   I   d.   Fe

      b.   Si   e.   Cu

      c.   W   f.   Co

100. a.   Br   d.   V

      b.   Bi   e.   F

      c.   Hg   f.   Ca

102. a.   osmium   e.   uranium

      b.   zirconium   f.   manganese

      c.   rubidium   g.   nickel

      d.   radon   h.   bromine

104. a. CO_2

      b. $AlCl_3$

      c. $HClO_4$

      d. SCl_6

106. a. $^{13}_{6}C$

      b. $^{13}_{6}C$

      c. $^{13}_{6}C$

      d. $^{44}_{19}K$

      e. $^{41}_{20}Ca$

      f. $^{35}_{19}K$

108.

Symbol	Number of Protons	Number of Neutrons	Mass Number
a. $^{41}_{20}\text{Ca}$	20	21	41
b. $^{55}_{25}\text{Mn}$	25	30	55
c. $^{109}_{47}\text{Ag}$	47	62	109
d. $^{45}_{21}\text{Sc}$	21	24	45

Chapter 5   Nomenclature

2.   A binary compound contains only two elements: the major types of binary compounds are *ionic* (compounds that contain a metal and a nonmetal) and *nonionic* (compounds containing two nonmetals).

4.   cation

6.   Some substances do not contain molecules. For example, the substance sodium chloride consists of an extended lattice array of sodium ions, Na^+, and chloride ions, Cl^-. Each sodium ion is surrounded by several chloride ions, and each chloride ion is surrounded by several sodium ions. We write the formula as NaCl to indicate the relative number of each ion in the substance, not to indicate that there are "molecules" of sodium chloride.

8.   Roman numeral

10.   a.   sodium bromide

     b.   calcium sulfide

     c.   aluminum iodide

     d.   cesium oxide

     e.   magnesium fluoride

     f.   strontium chloride

     g.   lithium oxide

     h.   barium iodide

12.   a.   incorrect; silver oxide is Ag_2O

     b.   correct

     c.   incorrect; barium oxide is BaO

     d.   correct

     e.   incorrect; lithium bromide is LiBr

14.   a.   As the oxide ion has a 2– charge, the tin ion must have a 2+ charge: the name is tin(II) oxide.

     b.   As each oxide ion has a 2– charge, the tin ion must have a 4+ charge: the name is tin(IV) oxide.

     c.   As the sulfide ion has a 2– charge, the copper ion must have a 2+ charge: the name is copper(II) sulfide.

     d.   As the sulfide ion has a 2– charge, the copper atoms must each have a 1+ charge: the name is copper(I) sulfide.

     e.   As each chloride ion has a 1– charge, the iron ion must have a 3+ charge: the name is iron(III) chloride.

f.   As each chloride ion has a 1- charge, the iron ion must have a 2+ charge: the name is iron(II) chloride.

16.   a.   As sulfide ions have a 2- charge, the tin ion must have a 2+ charge: the name is stan*nous* sulfide.

b.   As sulfide ions have a 2- charge, the tin ion must have a 4+ charge: the name is stan*nic* bromide.

c.   As the chloride ions each have a 1- charge, the cobalt ion must have a 3+ charge: the name is cobalt*ic* chloride.

d.   As the iodide ions each have a 1- charge, the cobalt ion must have a 2+ charge: the name is cobalt*ous* iodide.

e.   As the bromide ions each have a 1- charge, the mercury ion must have a 2+ charge: the name is mercur*ic* bromide.

f.   As the bromide ions each have a 1- charge, each mercury ion must have a 1+ charge: the name is mercur*ous* bromide.

18.   a.   xenon hexafluoride

b.   silicon tetrachloride

c.   diphosphorus pent(a)oxide

d.   water (dihydrogen oxide)

e.   arsenic tribromide

f.   carbon tetraiodide

20.   a.   diboron hexahydride - nonionic (common name: *diborane*)

b.   calcium nitride - ionic

c.   carbon tetrabromide - nonionic

d.   silver sulfide - ionic

e.   copper(II) chloride, cupric chloride - ionic

f.   chlorine monofluoride - nonionic

22.   a.   barium fluoride - ionic

b.   radium oxide - ionic

c.   dinitrogen oxide - nonionic

d.   rubidium oxide - ionic

e.   diarsenic pent(a)oxide - nonionic

f.   calcium nitride - ionic

24.   An oxyanion is a polyatomic anion containing oxygen combined with
       another element. The following oxyanions of bromine illustrate the
       nomenclature

       BrO^- hypobromite

       BrO_2^- bromite

       BrO_3^- bromate

       BrO_4^- perbromate

26.   For a series of oxyanions, the prefix *hypo-* is used for the anion with
       the fewest oxygen atoms, and the prefix *per-* is used for the anion with
       the most oxygen atoms.

28. BrO^- hypobromite

       BrO_2^- bromite

       BrO_3^- bromate

       BrO_4^- perbromate

30.   a. NO_3^-

       b. NO_2^-

       c. NH_4^+

       d. CN^-

32.   a. CO_3^{2-}

       b. HCO_3^-

       c. $C_2H_3O_2^-$

       d. CN^-

34.   a.   ammonium

       b.   dihydrogen phosphate

       c.   sulfate

       d.   hydrogen sulfite (also called *bi*sulfite)

       e.   perchlorate

       f.   iodate

36.   a.   ammonium sulfate

       b.   potassium perchlorate

       c.   iron(III) sulfate, ferric sulfate

       d.   calcium phosphate

  e. calcium hydroxide

  f. potassium carbonate

38. oxygen (commonly referred to as *oxyacids*)

40. a. hypochlorous acid

  b. sulfurous acid

  c. bromic acid

  d. hypoiodous acid

  e. perbromic acid

  f. hydrosulfuric acid

  g. hydroselenic acid

  h. phosphorous acid

42. a. PbO_2

  b. $SnBr_2$

  c. CuS

  d. CuI

  e. Hg_2Cl_2

  f. CrF_3

44. a. N_2O

  b. NO_2

  c. N_2O_4

  d. SF_6

  e. PBr_3

  f. CI_4

  g. OCl_2

46. a. $BaSO_3$

  b. $Ca(H_2PO_4)_2$

  c. NH_4ClO_4

  d. $NaMnO_4$

  e. $Fe_2(SO_4)_3$

  f. $CoCO_3$

g. $Ni(OH)_2$

h. $ZnCrO_4$

48.   a.   HCN

b. HNO_3

c. H_2SO_4

d. H_3PO_4

e.   HClO or HOCl

f.   HBr

g. $HBrO_2$

h.   HF

50.   a. $Mg(HSO_4)_2$

b. $CsClO_4$

c.   FeO

d. H_2Te

e. $Sr(NO_3)_2$

f. $Sn(C_2H_3O_2)_4$

g. $MnSO_4$

h. N_2O_4

i. Na_2HPO_4

j. Li_2O_2

k. HNO_2

l. $Co(NO_3)_3$

52.   A moist paste of NaCl would contain Na^+ and Cl^- ions in solution, and would serve as a *conductor* of electrical impulses.

54. $H \rightarrow H^+$ (hydrogen ion: a cation) $+ e^-$
$H + e^- \rightarrow H^-$ (*hydride* ion: an anion)

56.   missing oxyanions: IO_3^-; ClO_2^-

missing oxyacids: $HClO_4$; HClO; $HBrO_2$

58.   a.   gold(III) bromide, auric bromide

b.   cobalt(III) cyanide, cobaltic cyanide

c.   magnesium hydrogen phosphate

   d.   diboron hexahydride (diborane is its common name)

   e.   ammonia

   f.   silver(I) sulfate (usually called silver sulfate)

   g.   beryllium hydroxide

60.   a.   ammonium carbonate

   b.   ammonium hydrogen carbonate, ammonium bicarbonate

   c.   calcium phosphate

   d.   sulfurous acid

   e.   manganese(IV) oxide

   f.   iodic acid

   g.   potassium hydride

62.   a. $M(C_2H_3O_2)_4$ d. $M(HPO_4)_2$

   b. $M(MnO_4)_4$ e. $M(OH)_4$

   c. MO_2 f. $M(NO_2)_4$

64. M^+ compounds: MD, M_2E, M_3F

   M^{2+} compounds: MD_2, ME, M_3F_2

   M^{3+} compounds: MD_3, M_2E_3, MF

66.

$Ca(NO_3)_2$	$CaSO_4$	$Ca(HSO_4)_2$	$Ca(H_2PO_4)_2$	CaO	$CaCl_2$
$Sr(NO_3)_2$	$SrSO_4$	$Sr(HSO_4)_2$	$Sr(H_2PO_4)_2$	SrO	$SrCl_2$
NH_4NO_3	$(NH_4)_2SO_4$	NH_4HSO_4	$NH_4H_2PO_4$	$(NH_4)_2O$	NH_4Cl
$Al(NO_3)_3$	$Al_2(SO_4)_3$	$Al(HSO_4)_3$	$Al(H_2PO_4)_3$	Al_2O_3	$AlCl_3$
$Fe(NO_3)_3$	$Fe_2(SO_4)_3$	$Fe(HSO_4)_3$	$Fe(H_2PO_4)_3$	Fe_2O_3	$FeCl_3$
$Ni(NO_3)_2$	$NiSO_4$	$Ni(HSO_4)_2$	$Ni(H_2PO_4)_2$	NiO	$NiCl_2$
$AgNO_3$	Ag_2SO_4	$AgHSO_4$	AgH_2PO_4	Ag_2O	$AgCl$
$Au(NO_3)_3$	$Au_2(SO_4)_3$	$Au(HSO_4)_3$	$Au(H_2PO_4)_3$	Au_2O_3	$AuCl_3$
KNO_3	K_2SO_4	$KHSO_4$	KH_2PO_4	K_2O	KCl
$Hg(NO_3)_2$	$HgSO_4$	$Hg(HSO_4)_2$	$Hg(H_2PO_4)_2$	HgO	$HgCl_2$
$Ba(NO_3)_2$	$BaSO_4$	$Ba(HSO_4)_2$	$Ba(H_2PO_4)_2$	BaO	$BaCl_2$

68.   helium

70.   iodine (solid), bromine (liquid), fluorine and chlorine (gases)

72.   1-

74.   1–

76.   a. $Al(13e^-) \rightarrow Al^{3+}(10e^-) + 3e^-$

     b. $S(16e^-) + 2e^- \rightarrow S^{2-}(18e^-)$

     c. $Cu(29e^-) \rightarrow Cu^+(28e^-) + e^-$

     d. $F(9e^-) + e^- \rightarrow F^-(10e^-)$

     e. $Zn(30e^-) \rightarrow Zn^{2+}(28e^-) + 2e^-$

     f. $P(15e^-) + 3e^- \rightarrow P^{3-}(18e^-)$

78.   a.   Two 1+ ions are needed to balance a 2– ion, so the formula must have two Na^+ ions for each S^{2-} ion: Na_2S.

     b.   One 1+ ion exactly balances a 1– ion, so the formula should have an equal number of K^+ and Cl^- ions: KCl.

     c.   One 2+ ion exactly balances a 2– ion, so the formula must have an equal number of Ba^{2+} and O^{2-} ions: BaO.

     d.   One 2+ ion exactly balances a 2– ion, so the formula must have an equal number of Mg^{2+} and Se^{2-} ions: MgSe.

     e.   One 2+ ion requires two 1– ions to balance charge, so the formula must have twice as many Br^- ions as Cu^{2+} ions: $CuBr_2$.

     f.   One 3+ ion requires three 1– ions to balance charge, so the formula must have three times as many I^- ions as Al^{3+} ions: AlI_3.

     g.   Two 3+ ions give a total of 6+, whereas three 2– ions will give a total of 6–. The formula then should contain two Al^{3+} ions and three O^{2-} ions: Al_2O_3.

     h.   Three 2+ ions are required to balance two 3– ions, so the formula must contain three Ca^{2+} ions for every two N^{3-} ions: Ca_3N_2.

80.   a.   incorrect. Si is the element silicon, not silver.

     b.   incorrect. Co is the symbol for cobalt, not copper.

     c.   incorrect. Hydrogen exists as the hydride ion in this compound.

     d.   correct

     e.   incorrect. P is just "phosphorus" not "phosphoric".

82. a.   As bromide ions always have a 1- charge, the cobalt ion must have a 3+ charge: the name is cobalt*ic* bromide.

b.   As iodide ions always have a 1- charge, the lead ion must have a 4+ charge: the name is plumb*ic* iodide.

c.   As oxide ions always have a 2- charges, each iron ion must have a 3+ charge: the name is ferr*ic* oxide.

d.   As sulfide ions always have a 2- charge, the iron ion must have a 2+ charge: the name is ferr*ous* sulfide.

e.   As chloride ions always have a 1- charge, the tin ion must have a 4+ charge: the name is stann*ic* chloride.

f.   As oxide ions always have a 2- charge, the tin ion must have a 2+ charge: the name is stann*ous* oxide.

84. a.   iron(III) acetate, ferric acetate

b.   bromine monofluoride

c.   potassium peroxide

d.   silicon tetrabromide

e.   copper(II) permanganate, cupric permanganate

f.   calcium chromate

86. a. CO_3^{2-}

b. HCO_3^-

c. $C_2H_3O_2^-$

d. CN^-

88. a.   carbonate   d.   phosphate

b.   chlorate   e.   perchlorate

c.   sulfate   f.   permanganate

90. a. $CaCl_2$ e. H_2S

b. Ag_2O f. KH

c. Al_2S_3 g. MgI_2

d. $BeBr_2$ h. CsF

92. a. NaH_2PO_4 d. $KC_2H_3O_2$

b. $LiClO_4$ e. BaO_2

c. $Cu(HCO_3)_2$ f. Cs_2SO_3

Cumulative Review: Chapters 4 and 5

2.   How many elements could you name? Although you certainly don't have to memorize all the elements, you should at least be able to give the symbol or name for the most common elements (listed in Table 4.3).

4.   Dalton's atomic theory as presented in this text consists of five main postulates. Although Dalton's theory was exceptional scientific thinking for its time, some of the postulates have been modified as our scientific instruments and calculation methods have become increasingly more sophisticated. The main postulates of Dalton's theory are as follows: (1) Elements are made up of tiny particles called atoms; (2) all atoms of a given element are identical; (3) although all atoms of a given element are identical, these atoms are different from the atoms of all other elements; (4) atoms of one element can combine with atoms of another element to form a compound, and such a compound will always have the same relative numbers and types of atoms for its composition; and (5) atoms are merely rearranged into new groupings during an ordinary chemical reaction, and no atom is ever destroyed and no new atom is ever created during such a reaction.

6.   The expression *nuclear* atom indicates that we view the atom as having a dense center of positive charge (called the nucleus) around which the electrons move through primarily empty space. Rutherford's experiment involved shooting a beam of particles at a thin sheet of metal foil. According to the then current "plum pudding" model of the atom, most of these positively-charged particles should have passed right through the foil. However, Rutherford detected that a significant number of particles effectively bounced off something and were deflected backwards to the source of particles, and that other particles were deflected from the foil at large angles. Rutherford realized that his observations could be explained if the atoms of the metal foil had a small, dense, positively-charged nucleus, with a significant amount of empty space between nuclei. The empty space between nuclei would allow most of the particles to pass through the atom. However, if a particle hit a nucleus head-on, it would be deflected backwards at the source. If a positively-charged particle passed near a positively-charged nucleus (but did not hit the nucleus head-on), then the particle would be deflected by the repulsive forces between the positive charges. Rutherford's experiment conclusively disproved the "plum pudding" model for the atom, which envisioned the atom as a uniform sphere of positive charge, with enough negativel-charged electrons scattered through the atom to balance out the positive charge.

8.   Isotopes represent atoms of the same element which have different atomic masses. Isotopes are a result of the fact that atoms of a given element may have different numbers of neutrons in their nuclei. Isotopes have the same atomic number (number of protons in the nucleus) but have different mass numbers (total number of protons and neutrons in the nucleus). The different isotopes of an atom are indicated by symbolism of the form $_Z^A X$ in which Z represents the atomic number, and A the mass number, of element X. For example, $_6^{13}C$ represents a nuclide of carbon

with atomic number 6 (6 protons in the nucleus) and mass number 13 (reflecting 6 protons plus 7 neutrons in the nucleus). The various isotopes of an element have identical chemical properties because the chemical properties of an atom are a function of the electrons in the atom (*not* the nucleus). The physical properties of the isotopes of an element (and compounds containing those isotopes) may differ because of the difference in mass of the isotopes.

10. Most elements are too reactive to be found in nature in other than the combined form. Aside from the noble metals gold, silver, and platinum, the only other elements commonly found in nature in the uncombined state are some of the gaseous elements (such as O_2, N_2, He, Ar, etc.), and the solid nonmetals carbon and sulfur.

12. Ionic compounds typically are hard, crystalline solids with high melting and boiling points. Ionic substances like sodium chloride, when dissolved in water or when melted, conduct electrical currents: chemists have taken this evidence to mean that ionic substances consist of positively- and negatively-charged particles (ions). Although an ionic substance is made up of positively- and negatively-charged particles, there is no net electrical charge on a sample of such a substance because the total number of positive charges is balanced by an equal number of negative charges. An ionic compound could not possibly exist of just cations or just anions: there must be a balance of charge or the compound would be very unstable (like charges repel each other).

14. When naming ionic compounds, we name the positive ion (cation) first. For simple binary Type I ionic compounds, the ending -*ide* is added to the root name of the element that is the negative ion (anion). For example, for the Type I ionic compound formed between potassium and sulfur, K_2S, the name would be potassium sulfide: potassium is the cation, sulfur is the anion (with the suffix -*ide* added). Type II compounds are named by either of two systems, the "*ous-ic*" system (which is falling out of use), and the "Roman numeral" system which is preferred by most chemists. Type II compounds involve elements that form more than one stable ion. It is therefore necessary to specify *which* ion is present in a given compound. For example, iron forms two types of stable ion: Fe^{2+} and Fe^{3+}. Iron can react with oxygen to form either of two stable oxides, FeO or Fe_2O_3, depending on which cation is involved. Under the Roman numeral naming system, FeO would be named iron(II) oxide to show that it contains Fe^{2+} ions; Fe_2O_3 would be named iron(III) oxide to indicate that it contains Fe^{3+} ions. The Roman numeral used in a name corresponds to the charge of the specific ion present in the compound. Under the less-favored "*ous-ic*" system, for an element that forms two stable ions, the ending -*ous* is used to indicate the lower-charged ion, whereas the ending -*ic* is used to indicate the higher-charged ion. FeO and Fe_2O_3 would thus be named ferr*ous* oxide and ferr*ic* oxide, respectively. The "*ous-ic*" system has fallen out of favor because it does not indicate the actual charge on the ion, but only that it is the lower or higher charged of the two. This can lead to confusion: for example Fe^{2+} is called ferrous ion in this system, but Cu^{2+} is called cupric ion (since there is also a Cu^+ stable ion).

16. A polyatomic ion is an ion containing more than one atom. Some common polyatomic ions you should be familiar with are listed in Table 5.4. Parentheses are used in writing formulas containing polyatomic ions to indicate unambiguously how many of the polyatomic ion are present in the formula, to make certain that there is no mistake as to what is meant by the formula. For example, consider the substance calcium phosphate. The correct formula for this substance is $Ca_3(PO_4)_2$, which indicates that three calcium ions are combined for every two phosphate ions (check the total number of positive and negative charges to see why this is so). If we did not write the parenthesis around the formula for the phosphate ion, that is, if we had written Ca_3PO_{42}, people reading this formula might think that there were 42 oxygen atoms present!

18. Acids, in general, are substances that produce protons (H^+ ions) when dissolved in water. For acids that do not contain oxygen, the prefix *hydro-* and the suffix *-ic* are used with the root name of the element present in the acid (for example: HCl, hydrochloric acid; H_2S, hydrosulfuric acid; HF, hydrofluoric acid). The nomenclature of acids whose anions contain oxygen is more complicated. A series of prefixes and suffixes is used with the name of the non-oxygen atom in the anion of the acid: these prefixes and suffixes indicate the relative (not actual) number of oxygen atoms present in the anion. Most of the elements that form oxyanions form two such anions: for example, sulfur forms sulfite ion (SO_3^{2-}) and sulfate ion (SO_4^{2-}), and nitrogen forms nitrite ion (NO_2^-) and nitrate ion (NO_3^-). For an element that forms two oxyanions, the acid containing the anions will have the ending *-ous* if the anion is the *-ite* anion and the ending *-ic* if the anion is the *-ate* anion. For example, HNO_2 is nit*rous* acid and HNO_3 is nit*ric* acid; H_2SO_3 is sulfur*ous* acid and H_2SO_4 is sulfur*ic* acid. The halogen elements (Group 7) each form four oxyanions, and consequently, four oxyacids. The prefix *hypo-* is used for the oxyacid that contains fewer oxygen atoms than the *-ite* anion, and the prefix *per-* is used for the oxyacid that contains more oxygen atoms than the *-ate* anion. For example,

Acid	Name	Anion	Anion name
HBrO	*hypo*brom*ous* acid	BrO^-	*hypo*brom*ite*
$HBrO_2$	brom*ous* acid	BrO_2^-	brom*ite*
$HBrO_3$	brom*ic* acid	BrO_3^-	brom*ate*
$HBrO_4$	*per*brom*ic* acid	BrO_4^-	*per*brom*ate*

20.

	Symbol	Name	Atomic Number	Group Number
a.	Br	bromine	35	7A
b.	Li	lithum	3	1A
c.	Sr	strontium	38	2A
d.	K	potassium	19	1A
e.	B	boron	5	3A
f.	Cl	chlorine	17	7A
g.	Xe	xenon	54	8A
h.	C	carbon	6	4A
i.	Mg	magnesium	12	2A
j.	Al	aluminum	13	3A

22.

		Protons	Neutrons	Electrons
a.	$^{4}_{2}\text{He}$	2	2	2
b.	$^{37}_{17}\text{Cl}$	17	20	17
c.	$^{79}_{35}\text{Br}$	35	44	35
d.	$^{41}_{20}\text{Ca}$	20	21	20
e.	$^{40}_{20}\text{Ca}$	20	20	20
f.	$^{238}_{92}\text{U}$	92	146	92
g.	$^{235}_{92}\text{U}$	92	143	92
h.	$^{1}_{1}\text{H}$	1	0	1

24. a. 12 protons, 10 electrons

b. 26 protons, 24 electrons

c. 26 protons, 23 electrons

d. 9 protons, 10 electrons

e.   28 protons, 26 electrons

f.   30 protons, 28 electrons

g.   27 protons, 24 electrons

h.   7 protons, 10 electrons

i.   16 protons, 18 electrons

j.   37 protons, 36 electrons

k.   34 protons, 36 electrons

l.   19 protons, 18 electrons

26.   a. $FeCl_3$, iron(III) chloride, ferric chloride

b. Cu_2S, copper(I) sulfide, cuprous sulfide

c. $CoBr_2$, cobalt(II) bromide, cobaltous bromide

d. Fe_2O_3, iron(III) oxide, ferric oxide

e. AuI_3, gold(III) iodide, auric iodide

f. Cr_2S_3, chromium(III) sulfide, chromic sulfide

g. MnO_2, manganese(IV) oxide, manganese dioxide (archaic)

h.   CuO, copper(II) oxide, cupric oxide

i.   NiS, nickel(II) sulfide, nickelous sulfide

28.   a. NH_4^+, ammonium ion   f. CN^-, cyanide ion

b. SO_3^{2-}, sulfite ion   g. OH^-, hydroxide ion

c. NO_3^-, nitrate ion   h. ClO_4^-, perchlorate ion

d. SO_4^{2-}, sulfate ion   i. ClO^-, hypochlorite ion

e. NO_2^-, nitrite ion   j. PO_4^{3-}, phosphate ion

30.   a. B_2O_3, diboron trioxide

b. NO_2, nitrogen dioxide

c. PCl_5, phosphorus pentachloride

d. N_2O_4, dinitrogen tetroxide

e. P_2O_5, diphosphorus pentoxide

f.   ICl, iodine monochloride

g. SF_6, sulfur hexafluoride

h. N_2O_3, dinitrogen trioxide

Chapter 6   Chemical Reactions: An Introduction

2. The air pockets that form inside the biscuit are a result of the evolution of CO_2 gas by the baking powder.

4. Bubbling takes place as the hydrogen peroxide chemically decomposes into water and oxygen gas.

6. The alcohol in the wine is converted by wild yeasts in the air into acetic acid (vinegar). The observation that the odor of the wine has changed to the odor of vinegar indicates that a new substance has been produced. This is a chemical reaction.

8. atoms

10. the same

12. water

14. $C_3H_8(g) + O_2(g) \rightarrow CO_2(g) + H_2O(g)$

16. $BaCl_2(aq) + Na_2SO_4(aq) \rightarrow BaSO_4(s) + NaCl(aq)$

18. $C_3H_8(g) + O_2(g) \rightarrow CO_2(g) + H_2O(g)$
    $C_3H_3(g) + O_2(g) \rightarrow CO(g) + H_2O(g)$

20. $CaCO_3(s) + HCl(aq) \rightarrow CaCl_2(aq) + H_2O(l) + CO_2(g)$

22. $SiO_2(s) + C(s) \rightarrow Si(s) + CO(g)$

24. $H_2S(g) + O_2(g) \rightarrow SO_2(g) + H_2O(g)$

26. $Fe_2O_3(s) + CO(g) \rightarrow Fe(l) + CO_2(g)$

28. $O_2(g) \rightarrow O_3(g)$

30. $NH_3(g) + HNO_3(aq) \rightarrow NH_4NO_3(s)$

32. $Xe(g) + F_2(g) \rightarrow XeF_4(s)$

34. $BaO_2(s) + H_2SO_4(aq) \rightarrow BaSO_4(s) + H_2O_2(aq)$

36. whole numbers

38. a. $H_2O_2 \rightarrow H_2O + O_2$

     Balance oxygen: $2H_2O_2 \rightarrow 2H_2O + O_2$

     Balanced equation: $2H_2O_2(aq) \rightarrow 2H_2O(l) + O_2(g)$

b. $Ag + H_2S \rightarrow Ag_2S + H_2$

Balance silver: **2**$Ag + H_2S \rightarrow Ag_2S + H_2$

Balanced equation: $2Ag(s) + H_2S(g) \rightarrow Ag_2S(s) + H_2(g)$

c. $FeO + C \rightarrow Fe + CO_2$

Balance oxygen: **2**$FeO + C \rightarrow Fe + CO_2$

Balance iron: $2FeO + C \rightarrow$ **2**$Fe + CO_2$

Balanced equation: $2FeO(s) + C(s) \rightarrow 2Fe(l) + CO_2(g)$

d. $Cl_2 + KI \rightarrow KCl + I_2$

Balance chlorine: $Cl_2 + KI \rightarrow$ **2**$KCl + I_2$

Balance iodine: $Cl_2 +$ **2**$KI \rightarrow 2KCl + I_2$

Balanced equation: $Cl_2(g) + 2KI(aq) \rightarrow 2KCl(aq) + I_2(s)$

e. $Na_2B_4O_7 + H_2SO_4 + H_2O \rightarrow H_3BO_3 + Na_2SO_4$

Balance boron: $Na_2B_4O_7 + H_2SO_4 + H_2O \rightarrow$ **4**$H_3BO_3 + Na_2SO_4$

Balance hydrogen: $Na_2B_4O_7 + H_2SO_4 +$ **5**$H_2O \rightarrow 4H_3BO_3 + Na_2SO_4$

Balanced equation: $Na_2B_4O_7(s) + H_2SO_4(aq) + 5H_2O(l)$

$$\rightarrow 4H_3BO_3(s) + Na_2SO_4(aq)$$

f. $CaC_2 + H_2O \rightarrow Ca(OH)_2 + C_2H_2$

Balance oxygen: $CaC_2 +$ **2**$H_2O \rightarrow Ca(OH)_2 + C_2H_2$

Balanced equation: $CaC_2(s) + 2H_2O(l) \rightarrow Ca(OH)_2(s) + C_2H_2(g)$

g. $NaCl + H_2SO_4 \rightarrow HCl + Na_2SO_4$

Balance sodium: **2**$NaCl + H_2SO_4 \rightarrow HCl + Na_2SO_4$

Balance chlorine: $2NaCl + H_2SO_4 \rightarrow$ **2**$HCl + Na_2SO_4$

Balanced equation: $2NaCl(s) + H_2SO_4(l) \rightarrow 2HCl(g) + Na_2SO_4(s)$

h. $SiO_2 + C \rightarrow Si + CO$

Balance oxygen: $SiO_2 + C \rightarrow Si +$ **2**CO

Balance carbon: $SiO_2 +$ **2**$C \rightarrow Si + 2CO$

Balanced equation: $SiO_2(s) + 2C(s) \rightarrow Si(l) + 2CO(g)$

40.   a. $CaF_2 + H_2SO_4 \rightarrow CaSO_4 + HF$

Balance fluorine: $CaF_2 + H_2SO_4 \rightarrow CaSO_4 +$ **2**HF

Balanced equation: $CaF_2(s) + H_2SO_4(l) \rightarrow CaSO_4(s) + 2HF(g)$

b. $KBr + H_3PO_4 \rightarrow K_3PO_4 + HBr$

Balance potassium: **3**$KBr + H_3PO_4 \rightarrow K_3PO_4 + HBr$

Balance bromine: $3KBr + H_3PO_4 \rightarrow K_3PO_4 +$ **3**HBr

Balanced equation: $3KBr(s) + H_3PO_4(aq) \rightarrow K_3PO_4(aq) + 3HBr(g)$

c. $TiCl_4$ + Na → NaCl + Ti

Balance chlorine: $TiCl_4$ + Na → **4**NaCl + Ti

Balance sodium: $TiCl_4$ + **4**Na → 4NaCl + Ti

Balanced equation: $TiCl_4$(*l*) + 4Na(*s*) → 4NaCl(*s*) + Ti(*s*)

d. K_2CO_3 → K_2O + CO_2 This equation is already balanced!

e. KO_2 + H_2O → KOH + O_2

Balance hydrogen: KO_2 + H_2O → **2**KOH + O_2

Balance potassium: **2**KO_2 + H_2O → 2KOH + O_2

At this point, we have balanced potassium and hydrogen atoms, but now it becomes difficult to balance oxygen because it occurs in each of the reactants and products. There is no systematic way to balance oxygen at this point (you will learn a special method in a later chapter for oxidation-reduction reactions such as this). We need one more oxygen atom on the right side of the equation to balance it: if we could just have an extra half of an O_2 molecule (that is, $1.5O_2$) the equation would be balanced. That is

Balance using non-integer: $2KO_2$ + H_2O → 2KOH + **1.5**O_2

Although this equation is balanced, we can't really have 1.5 molecules. If we multiply everything in this equation by 2, however, we will get whole number coefficients.

Balanced equation: $4KO_2$(*s*) + $2H_2O$(*l*) → 4KOH(*aq*) + $3O_2$(*g*)

f. Na_2O_2 + H_2O + CO_2 → $NaHCO_3$ + O_2

Balance sodium: Na_2O_2 + H_2O + CO_2 → **2**$NaHCO_3$ + O_2

Balance carbon: Na_2O_2 + H_2O + **2**CO_2 → 2$NaHCO_3$ + O_2

Again we are left with the difficulty of balancing oxygen when it occurs in all the reactants and products. Right now, there are 7 oxygen atoms on the left side of the equation, but there are 8 oxygen atoms on the right side. If we could have just half an O_2 molecule (instead of one entire molecule) things would work out.

Balance using non-integer: Na_2O_2 + H_2O + 2CO_2 → 2$NaHCO_3$ + **0.5**O_2

Although this equation is balanced, we can't really have half a molecule. If we multiply everything in the equation by 2, however, we'll get whole numbers.

Balanced: $2Na_2O_2$(*s*) + $2H_2O$(*g*) + $4CO_2$(*g*) → 4$NaHCO_3$(*s*) + O_2(*g*)

g. KNO_2 + C → K_2CO_3 + CO + N_2

Balance nitrogen: **2**KNO_2 + C → K_2CO_3 + CO + N_2

Balance carbon: 2KNO_2 + **2**C → K_2CO_3 + CO + N_2

Balanced equation: $2KNO_2$(*s*) + 2C(*s*) → K_2CO_3(*s*) + CO(*g*) + N_2(*g*)

h. $BaO + Al \rightarrow Ba + Al_2O_3$

Balance aluminum: $BaO + \mathbf{2}Al \rightarrow Ba + Al_2O_3$

Balance oxygen: $\mathbf{3}BaO + 2Al \rightarrow Ba + Al_2O_3$

Balance barium: $3BaO + 2Al \rightarrow \mathbf{3}Ba + Al_2O_3$

Balanced equation: $3BaO(s) + 2Al(s) \rightarrow 3Ba(s) + Al_2O_3(s)$

42. a. $2Ag_2S(s) + 2H_2O(l) \rightarrow 4Ag(s) + 2H_2S(g) + O_2(g)$

b. $CaO(s) + SO_3(s) \rightarrow CaSO_4(s)$

c. $CS_2(g) + 3Cl_2(g) \rightarrow CCl_4(l) + S_2Cl_2(g)$

d. $2Al(s) + 3F_2(g) \rightarrow 2AlF_3(s)$

e. $4NH_3(g) + 5O_2(g) \rightarrow 4NO(g) + 6H_2O(l)$

f. $PI_3(s) + 3H_2O(l) \rightarrow H_3PO_3(aq) + 3HI(g)$

g. $4C(s) + S_8(s) \rightarrow 4CS_2(l)$

h. $CaH_2(s) + 2H_2O(l) \rightarrow Ca(OH)_2(aq) + 2H_2(g)$

44. a. $Ba(NO_3)_2(aq) + Na_2CrO_4(aq) \rightarrow BaCrO_4(s) + 2NaNO_3(aq)$

b. $PbCl_2(aq) + K_2SO_4(aq) \rightarrow PbSO_4(s) + 2KCl(aq)$

c. $C_2H_5OH(l) + 3O_2(g) \rightarrow 2CO_2(g) + 3H_2O(l)$

d. $CaC_2(s) + 2H_2O(l) \rightarrow Ca(OH)_2(s) + C_2H_2(g)$

e. $Sr(s) + 2HNO_3(aq) \rightarrow Sr(NO_3)_2(aq) + H_2(g)$

f. $BaO_2(s) + H_2SO_4(aq) \rightarrow BaSO_4(s) + H_2O_2(aq)$

g. $2AsI_3(s) \rightarrow 2As(s) + 3I_2(s)$

h. $2CuSO_4(aq) + 4KI(s) \rightarrow 2CuI(s) + I_2(s) + 2K_2SO_4(aq)$

46. $Al(s) + O_2(g) \rightarrow Al_2O_3(s)$

48. $C_{12}H_{22}O_{11}(aq) + H_2O(l) \rightarrow 4C_2H_5OH(aq) + 4CO_2(g)$

50. $2Al_2O_3(s) + 3C(s) \rightarrow 4Al(s) + 3CO_2(g)$

52. $2Li(s) + S(s) \rightarrow Li_2S(s)$ $2Rb(s) + S(s) \rightarrow Rb_2S(s)$

$2Cs(s) + S(s) \rightarrow Cs_2S(s)$ $2Na(s) + S(s) \rightarrow Na_2S(s)$

$2Fr(s) + S(s) \rightarrow Fr_2S(s)$ $2K(s) + S(s) \rightarrow K_2S(s)$

54. $BaO_2(s) + H_2O(l) \rightarrow BaO(s) + H_2O_2(aq)$

56. $2KClO_3(s) \rightarrow 2KCl(s) + 3O_2(g)$

58. $NH_3(g) + HCl(g) \rightarrow NH_4Cl(s)$

60.   The "charring" represents the conversion of the carbohydrates (starch) in the muffin to elemental carbon.

62. $Fe(s) + S(s) \rightarrow FeS(s)$

64. $K_2CrO_4(aq) + BaCl_2(aq) \rightarrow BaCrO_4(s) + 2KCl(aq)$

66. $2NaCl(aq) + 2H_2O(l) \rightarrow 2NaOH(aq) + H_2(g) + Cl_2(g)$

$2NaBr(aq) + 2H_2O(l) \rightarrow 2NaOH(aq) + H_2(g) + Br_2(g)$

$2NaI(aq) + 2H_2O(l) \rightarrow 2NaOH(aq) + H_2(g) + I_2(g)$

68. $CaC_2(s) + 2H_2O(l) \rightarrow Ca(OH)_2(s) + C_2H_2(g)$

70. $CuO(s) + H_2SO_4(aq) \rightarrow CuSO_4(aq) + H_2O(l)$

72. $Na_2SO_3(aq) + S(s) \rightarrow Na_2S_2O_3(aq)$

74.   a. $ZnCl_2(aq) + Na_2CO_3(aq) \rightarrow ZnCO_3(s) + 2NaCl(aq)$

b. $2Al(s) + 3H_2SO_4(aq) \rightarrow Al_2(SO_4)_3(aq) + 3H_2(g)$

c. $Mn(s) + 2S(s) \rightarrow MnS_2(s)$

d. $C_5H_{12}(l) + 8O_2(g) \rightarrow 5CO_2(g) + 6H_2O(g)$

e. $H_2O(l) + Br_2(l) \rightarrow HBr(aq) + HOBr(aq)$

f. $MnS_2(s) + 3O_2(g) \rightarrow MnO_2(s) + 2SO_2(g)$

g. $PbCl_2(aq) + K_2CrO_4(aq) \rightarrow PbCrO_4(s) + 2KCl(aq)$

h. $2AgNO_3(aq) + H_2SO_4(aq) \rightarrow Ag_2SO_4(s) + 2HNO_3(aq)$

76.   a. $Pb(NO_3)_2(aq) + K_2CrO_4(aq) \rightarrow PbCrO_4(s) + 2KNO_3(aq)$

b. $BaCl_2(aq) + Na_2SO_4(aq) \rightarrow BaSO_4(s) + 2NaCl(aq)$

c. $2CH_3OH(l) + 3O_2(g) \rightarrow 2CO_2(g) + 4H_2O(g)$

d. $Na_2CO_3(aq) + S(s) + SO_2(g) \rightarrow CO_2(g) + Na_2S_2O_3(aq)$

e. $Cu(s) + 2H_2SO_4(aq) \rightarrow CuSO_4(aq) + SO_2(g) + 2H_2O(l)$

f. $MnO_2(s) + 4HCl(aq) \rightarrow MnCl_2(aq) + Cl_2(g) + 2H_2O(l)$

g. $As_2O_3(s) + 6KI(aq) + 6HCl(aq) \rightarrow 2AsI_3(s) + 6KCl(aq) + 3H_2O(l)$

h. $2Na_2S_2O_3(aq) + I_2(aq) \rightarrow Na_2S_4O_6(aq) + 2NaI(aq)$

Chapter 7 Reactions in Aqueous Solutions

2. Driving forces are types of *changes* in a system that pull a reaction in the *direction of product formation*; driving forces discussed in Chapter 7 include: formation of a *solid*, formation of *water*, formation of a *gas*, and transfer of electrons.

4. The net charge of a precipitate must be *zero*. The total number of positive charges equals the total number of negative charges.

6. ions

8. Chemists know that a solution contains independent ions because such a solution will readily allow an electrical current to pass through it. The simplest experiment that demonstrates this uses the sort of light-bulb conductivity apparatus described in the text: if the light bulb glows strongly, then the solution must contain a lot of ions to be conducting the electricity well.

10. For most practical purposes, "insoluble" and "slightly" soluble mean the same thing. The difference between "insoluble" and "slightly soluble" could be crucial if, for example, a substance were highly toxic and were found in a water supply.

12. a. soluble (Rule 3: Most chloride salts are soluble.)

    b. soluble (Rule 1: Most nitrate salts are soluble.)

    c. insoluble (Rule 4: exception for sulfate salts)

    d. soluble (Rule 1: Most sodium salts are soluble.)

    e. insoluble (Rule 6: Most sulfide salts are insoluble.)

    f. insoluble (Rule 5: Most hydroxide compounds are insoluble.)

    g. insoluble (Rule 6: Most carbonate salts are insoluble.)

    h. soluble (Rule 1: Most ammonium salts are soluble.)

14. a. Rule 4: exception to the rule for sulfates

    b. Rule 6: Most sulfide salts are only slightly soluble.

    c. Rule 5: Most hydroxide compounds are only slightly soluble.

    d. Rule 6: Most carbonate compounds are only slightly soluble.

16. a. $CaSO_4$. Rule 4: exception to the rule for sulfates

    b. AgI. Rule 3: Although the text does not mention it explicitly, as you might expect from your knowledge of the periodic table, bromide and iodide compounds of Ag^+, Pb^{2+}, and Hg_2^{2+} are insoluble.

    c. $Pb_3(PO_4)_2$. Rule 6: Most phosphate salts are only slightly soluble.

    d. $Fe(OH)_3$. Rule 5: Most hydroxides are only slightly soluble.

e.   No precipitate is likely: Rules 1, 2, and 4

f. $BaCO_3$. Rule 6: Most carbonate salts are only slightly soluble.

18.   The precipitates are marked in boldface type.

a.   Rule 6: Most sulfide salts are insoluble.

$Na_2S(aq) + CuCl_2(aq) \rightarrow$ **$CuS(s)$** $+ 2NaCl(aq)$

b.   Rule 6: Most phosphate salts are insoluble.

$K_3PO_4(aq) + AlCl_3(aq) \rightarrow 3KCl(aq) +$ **$AlPO_4(s)$**

c.   Rule 4: barium sulfate is a listed exception.

$H_2SO_4(aq) + BaCl_2(aq) \rightarrow$ **$BaSO_4(s)$** $+ 2HCl(aq)$

d.   Rule 5: Most hydroxide compounds are insoluble.

$3NaOH(aq) + FeCl_3(aq) \rightarrow 3NaCl(aq) +$ **$Fe(OH)_3(s)$**

e.   Rule 3: exception to the rule for chlorides

$2NaCl(aq) + Hg_2(NO_3)_2(aq) \rightarrow 2NaNO_3(aq) +$ **$Hg_2Cl_2(s)$**

f.   Rule 6: Most carbonate salts are insoluble.

$3K_2CO_3(aq) + 2Cr(C_2H_3O_2)_3(aq) \rightarrow 6KC_2H_3O_2(aq) +$ **$Cr_2(CO_3)_3(s)$**

20.   Hint: when balancing equations involving polyatomic ions, especially in precipitation reactions, balance the polyatomic ions as a *unit*, not in terms of the atoms the polyatomic ions contain (e.g., treat nitrate ion, NO_3^- as a single entity, not as one nitrogen and three oxygen atoms). When finished balancing, however, be sure to count the individual number of atoms of each type on each side of the equation.

a. $AgNO_3(aq) + H_2SO_4(aq) \rightarrow Ag_2SO_4(s) + HNO_3(aq)$

Balance silver: **2**$AgNO_3(aq) + H_2SO_4(aq) \rightarrow Ag_2SO_4(s) + HNO_3(aq)$

Balance nitrate: $2AgNO_3(aq) + H_2SO_4(aq) \rightarrow Ag_2SO_4(s) +$ **2**$HNO_3(aq)$

Balanced equation: $2AgNO_3(aq) + H_2SO_4(aq) \rightarrow Ag_2SO_4(s) + 2HNO_3(aq)$

b. $Ca(NO_3)_2(aq) + H_2SO_4(aq) \rightarrow CaSO_4(s) + HNO_3(aq)$

Balance nitrate: $Ca(NO_3)_2(aq) + H_2SO_4(aq) \rightarrow CaSO_4(s) +$ **2**$HNO_3(aq)$

Balanced equation: $Ca(NO_3)_2(aq) + H_2SO_4(aq) \rightarrow CaSO_4(s) + 2HNO_3(aq)$

c. $Pb(NO_3)_2(aq) + H_2SO_4(aq) \rightarrow PbSO_4(s) + HNO_3(aq)$

Balance nitrate: $Pb(NO_3)_2(aq) + H_2SO_4(aq) \rightarrow PbSO_4(s) +$ **2**$HNO_3(aq)$

Balanced equation: $Pb(NO_3)_2(aq) + H_2SO_4(aq) \rightarrow PbSO_4(s) + 2HNO_3(aq)$

22. The products are determined by having the ions "switch partners." For example, for a general reaction AB + CD →, the possible products are AD and CB if the ions switch partners. If either AD or CB is insoluble, then a precipitation reaction has occurred. In the following reaction, the formula of the precipitate is given in boldface type.

   a. $(NH_4)_2S(aq) + CoCl_2(aq) \rightarrow$ **CoS**$(s) + 2NH_4Cl(aq)$

   Rule 6: Most sulfide salts are only slightly soluble.

   b. $FeCl_3(aq) + 3NaOH(aq) \rightarrow$ **Fe(OH)**$_3(s) + 3NaCl(aq)$

   Rule 5: Most hydroxide compounds are only slightly soluble.

   c. $CuSO_4(aq) + Na_2CO_3(aq) \rightarrow$ **CuCO**$_3(s) + Na_2SO_4(aq)$

   Rule 6: Most carbonate salts are only slightly soluble.

24. spectator

26. The net ionic equation for a reaction indicates *only those ions that go to form the precipitate* and does not show the spectator ions present in the solutes mixed. The identity of the precipitate is determined from the Solubility Rules (Table 7.1).

   a.   No reaction: all combinations of the ions are soluble.

   b. $S^{2-}(aq) + Ca^{2+}(aq) \rightarrow CaS(s)$

   Rule 6: Most sulfide salts are only slightly soluble.

   c. $Ag^+(aq) + OH^-(aq) \rightarrow AgOH(s)$

   Rule 5: Most hydroxide compounds are insoluble.

   d. $2Fe^{3+}(aq) + 3CO_3^{2-}(aq) \rightarrow Fe_2(CO_3)_3(s)$

   Rule 6: Most carbonate salts are only slightly soluble.

   e. $Al^{3+}(aq) + PO_4^{3-}(aq) \rightarrow AlPO_4(s)$

   Rule 6: Most phosphate salts are only slightly soluble.

   f.   No reaction: all combinations of the ions are soluble.

28. $Ag^+(aq) + Cl^-(aq) \rightarrow AgCl(s)$

   $Pb^{2+}(aq) + 2Cl^-(aq) \rightarrow PbCl_2(s)$

   $Hg_2^{2+}(aq) + 2Cl^-(aq) \rightarrow Hg_2Cl_2(s)$

30. $Co^{2+}(aq) + S^{2-}(aq) \rightarrow CoS(s)$

   $2Co^{3+}(aq) + 3S^{2-}(aq) \rightarrow Co_2S_3(s)$

   $Fe^{2+}(aq) + S^{2-}(aq) \rightarrow FeS(s)$

   $2Fe^{3+}(aq) + 3S^{2-}(aq) \rightarrow Fe_2S_3(s)$

32. Strong bases fully produce hydroxide ions when dissolved in water. The strong bases are also strong electrolytes.

34. acids: HCl, H_2SO_4, HNO_3, $HClO_4$, HBr

bases: NaOH, KOH, RbOH, CsOH

36. salt

38. $RbOH(s) \rightarrow Rb^+(aq) + OH^-(aq)$

$CsOH(s) \rightarrow Cs^+(aq) + OH^-(aq)$

40. In general, the salt formed in an aqueous acid-base reaction consists of the *positive ion of the base* involved in the reaction, combined with the *negative ion of the acid*. The hydrogen ion of the strong acid combines with the hydroxide ion of the strong base to produce water, which is the other product of the acid-base reactions.

a. $KOH(aq) + HCl(aq) \rightarrow H_2O(l) + KCl(aq)$

b. $NaOH(aq) + HClO_4(aq) \rightarrow H_2O(l) + NaClO_4(aq)$

c. $CsOH(aq) + HNO_3(aq) \rightarrow H_2O(l) + CsNO_3(aq)$

d. $2KOH(aq) + H_2SO_4(aq) \rightarrow 2H_2O(l) + K_2SO_4(aq)$

42. transfer

44. The metallic element *loses* electrons and the nonmetallic element *gains* electrons.

46. Each nitrogen atom would gain three electrons to become an N^{3-} ion. Each nitrogen atom would gain three electrons, which means that an N_2 molecule would gain six electrons.

48. $AlBr_3$ is made up of Al^{3+} ions and Br^- ions. Aluminum atoms each lose three electrons to become Al^{3+} ions. Bromine atoms each gain one electron to become Br^- ions (so each Br_2 molecule gains two electrons to become two Br^- ions).

50. a. $Fe(s) + S(s) \rightarrow Fe_2S_3(s)$

Balance iron: $\mathbf{2}Fe + S \rightarrow Fe_2S_3$

Balance sulfur: $2Fe + \mathbf{3}S \rightarrow Fe_2S_3$

Balanced equation: $2Fe(s) + 3S(s) \rightarrow Fe_2S_3(s)$

b. $Zn(s) + HNO_3(aq) \rightarrow Zn(NO_3)_2(aq) + H_2(g)$

Balance nitrate ions: $Zn + \mathbf{2}HNO_3 \rightarrow Zn(NO_3)_2 + H_2$

Balanced equation: $Zn(s) + 2HNO_3(aq) \rightarrow Zn(NO_3)_2(aq) + H_2(g)$

c. $Sn(s) + O_2(g) \rightarrow SnO(s)$

Balance oxygen: $Sn + O_2 \rightarrow \mathbf{2}SnO$

Balance tin: $\mathbf{2}Sn + O_2 \rightarrow 2SnO$

Balanced equation: $2Sn(s) + O_2(g) \rightarrow 2SnO(s)$

d. $K(s) + H_2(g) \rightarrow KH(s)$

Balance hydrogen: $K + H_2 \rightarrow \mathbf{2}KH$

Balance potassium: $\mathbf{2}K + H_2 \rightarrow 2KH$

Balanced equation: $2K(s) + H_2(g) \rightarrow 2KH(s)$

e. $Cs(s) + H_2O(l) \rightarrow CsOH(aq) + H_2(g)$

Balance hydrogen: $Cs + \mathbf{2}H_2O \rightarrow \mathbf{2}CsOH + H_2$

Balance cesium: $\mathbf{2}Cs + 2H_2O \rightarrow 2CsOH + H_2$

Balanced equation: $2Cs(s) + 2H_2O(l) \rightarrow 2CsOH(aq) + H_2(g)$

52.   examples of formation of water:

$$HCl(aq) + NaOH(aq) \rightarrow H_2O(l) + NaCl(aq)$$

$$H_2SO_4(aq) + 2KOH(aq) \rightarrow 2H_2O(l) + K_2SO_4(aq)$$

examples of formation of a gaseous product

$$Mg(s) + 2HCl(aq) \rightarrow MgCl_2(aq) + H_2(g)$$

$$2KClO_3(s) \rightarrow 2KCl(s) + 3O_2(g)$$

54.   For each reaction, the type of reaction is first identified, followed by some of the reasoning that leads to this choice (there may be more than one way in which you can recognize a particular type of reaction).

a.   oxidation-reduction (Oxygen changes from the combined state to the elemental state.)

b.   oxidation-reduction (Copper changes from the elemental to the combined state; hydrogen changes from the combined to the elemental state.)

c.   acid-base (H_2SO_4 is a strong acid and NaOH is a strong base; water and a salt are formed.)

d.   acid-base, precipitation (H_2SO_4 is a strong acid, and $Ba(OH)_2$ is a base; water and a salt are formed; an insoluble product forms.)

e.   precipitation (From the Solubility Rules of Table 7.1, AgCl is only slightly soluble.)

f.   precipitation (From the Solubility Rules of Table 7.1, $Cu(OH)_2$ is only slightly soluble.)

g.   oxidation-reduction (Chlorine and fluorine change from the elemental to the combined state.)

h.   oxidation-reduction (Oxygen changes from the elemental to the combined state.)

i.   acid-base (HNO_3 is a strong acid and $Ca(OH)_2$ is a strong base; a salt and water are formed.)

56.   oxidation-reduction

58.   A decomposition reaction is one in which a given compound is broken down into simpler compounds or constituent elements. The reactions

$$CaCO_3(s) \rightarrow CaO(s) + CO_2(g)$$

$$2HgO(s) \rightarrow 2Hg(l) + O_2(g)$$

both represent decomposition reactions. Such reactions often (but not necessarily always) may be classified in other ways. For example, the reaction of $HgO(s)$ is also an oxidation-reduction reaction.

60.   Compounds like those in parts b and c of this problem, containing only carbon and hydrogen, are called *hydrocarbons*. When a hydrocarbon is reacted with oxygen (O_2), the hydrocarbon is almost always converted to carbon dioxide and water vapor. Because water molecules contain an odd number of oxygen atoms, and O_2 contains an even number of oxygen atoms, it is often difficult to balance such equations. For this reason, it is simpler to balance the equation using fractional coefficients if necessary, and then to multiply by a factor that will give whole number coefficients for the final balanced equation.

a. $C_2H_5OH(l) + O_2(g) \rightarrow CO_2(g) + H_2O(g)$

Balance carbon: $C_2H_5OH(l) + O_2(g) \rightarrow \mathbf{2}CO_2(g) + H_2O(g)$

Balance hydrogen: $C_2H_5OH(l) + O_2(g) \rightarrow 2CO_2(g) + \mathbf{3}H_2O(g)$

Balance oxygen: $C_2H_5OH(l) + \mathbf{3}O_2(g) \rightarrow 2CO_2(g) + 3H_2O(g)$

Balanced equation: $C_2H_5OH(l) + 3O_2(g) \rightarrow 2CO_2(g) + 3H_2O(g)$

b. $C_6H_{14}(l) + O_2(g) \rightarrow CO_2(g) + H_2O(g)$

Balance carbon: $C_6H_{14}(l) + O_2(g) \rightarrow \mathbf{6}CO_2(g) + H_2O(g)$

Balance hydrogen: $C_6H_{14}(l) + O_2(g) \rightarrow 6CO_2(g) + \mathbf{7}H_2O(g)$

Balance oxygen: $C_6H_{14}(l) + \mathbf{(19/2)}O_2(g) \rightarrow 6CO_2(g) + 7H_2O(g)$

Balanced equation: $2C_6H_{14}(l) + 19O_2(g) \rightarrow 12CO_2(g) + 14H_2O(g)$

c. $C_6H_{12}(l) + O_2(g) \rightarrow CO_2(g) + H_2O(g)$

Balance carbon: $C_6H_{12}(l) + O_2(g) \rightarrow \mathbf{6}CO_2(g) + H_2O(g)$

Balance hydrogen: $C_6H_{12}(l) + O_2(g) \rightarrow 6CO_2(g) + \mathbf{6}H_2O(g)$

Balanced equation: $C_6H_{12}(l) + 9O_2(g) \rightarrow 6CO_2(g) + 6H_2O(g)$

62. A reaction in which small molecules or atoms combine to make a larger molecule is called a *synthesis* reaction. An example would be the synthesis of sodium chloride from the elements

$$2Na(s) + Cl_2(g) \rightarrow 2NaCl(s).$$

A reaction in which a molecule is broken down into simpler molecules or atoms is called a *decomposition* reaction. An example would be the decomposition of sodium hydrogen carbonate when heated.

$$2NaHCO_3(s) \rightarrow Na_2CO_3(s) + CO_2(g) + H_2O(g).$$

Specific examples will depend on the students' input.

64. a. $2Co(s) + 3S(s) \rightarrow Co_2S_3(s)$

    b. $2NO(g) + O_2(g) \rightarrow 2NO_2(g)$

    c. $FeO(s) + CO_2(g) \rightarrow FeCO_3(s)$

    d. $2Al(s) + 3F_2(g) \rightarrow 2AlF_3(s)$

    e. $2NH_3(g) + H_2CO_3(aq) \rightarrow (NH_4)_2CO_3(s)$

66. a. $2NI_3(s) \rightarrow N_2(g) + 3I_2(s)$

    b. $BaCO_3(s) \rightarrow BaO(s) + CO_2(g)$

    c. $C_6H_{12}O_6(s) \rightarrow 6C(s) + 6H_2O(g)$

    d. $Cu(NH_3)_4SO_4(s) \rightarrow CuSO_4(s) + 4NH_3(g)$

    e. $3NaN_3(s) \rightarrow Na_3N(s) + 4N_2(g)$

68. In several cases, the given ion may be precipitated by *many* reactants. The following are only three of the possible examples.

    a. Chloride ion would precipitate when treated with solutions containing silver ion, lead(II) ion, or mercury(I) ion.

    $Ag^+(aq) + Cl^-(aq) \rightarrow AgCl(s)$

    $Pb^{2+}(aq) + 2Cl^-(aq) \rightarrow PbCl_2(s)$

    $Hg_2^{2+}(aq) + 2Cl^-(aq) \rightarrow Hg_2Cl_2(s)$

    b. Calcium ion would precipitate when treated with solutions containing sulfate ion, carbonate ion, and phosphate ion.

    $Ca^{2+}(aq) + SO_4^{2-}(aq) \rightarrow CaSO_4(s)$

    $Ca^{2+}(aq) + CO_3^{2-}(aq) \rightarrow CaCO_3(s)$

    $3Ca^{2+}(aq) + 2PO_4^{3-}(aq) \rightarrow Ca_3(PO_4)_2(s)$

c.   Iron(III) ion would precipitate when treated with solutions containing hydroxide, sulfide, or carbonate ions.

$$Fe^{3+}(aq) + 3OH^-(aq) \rightarrow Fe(OH)_3(s)$$

$$2Fe^{3+}(aq) + 3S^{2-}(aq) \rightarrow Fe_2S_3(s)$$

$$2Fe^{3+}(aq) + 3CO_3^{2-}(aq) \rightarrow Fe_2(CO_3)_3(s)$$

d.   Sulfate ion would precipitate when treated with solutions containing barium ion, calcium ion, or lead(II) ion.

$$Ba^{2+}(aq) + SO_4^{2-}(aq) \rightarrow BaSO_4(s)$$

$$Ca^{2+}(aq) + SO_4^{2-}(aq) \rightarrow CaSO_4(s)$$

$$Pb^{2+}(aq) + SO_4^{2-}(aq) \rightarrow PbSO_4(s)$$

e.   Mercury(I) ion would precipitate when treated with solutions containing chloride ion, sulfide ion, or carbonate ion.

$$Hg_2^{2+}(aq) + 2Cl^-(aq) \rightarrow Hg_2Cl_2(s)$$

$$Hg_2^{2+}(aq) + S^{2-}(aq) \rightarrow Hg_2S(s)$$

$$Hg_2^{2+}(aq) + CO_3^{2-}(aq) \rightarrow Hg_2CO_3(s)$$

f.   Silver ion would precipitate when treated with solutions containing chloride ion, sulfide ion, or carbonate ion.

$$Ag^+(aq) + Cl^-(aq) \rightarrow AgCl(s)$$

$$2Ag^+(aq) + S^{2-}(aq) \rightarrow Ag_2S(s)$$

$$2Ag^+(aq) + CO_3^{2-}(aq) \rightarrow Ag_2CO_3(s)$$

70.   The formulas of the salts are indicated in boldface type.

a. $HNO_3(aq) + KOH(aq) \rightarrow H_2O(l) + $ **KNO_3**(aq)

b. $H_2SO_4(aq) + Ba(OH)_2(aq) \rightarrow 2H_2O(l) + $ **$BaSO_4$**(s)

c. $HClO_4(aq) + NaOH(aq) \rightarrow H_2O(l) + $ **$NaClO_4$**(aq)

d. $2HCl(aq) + Ca(OH)_2(aq) \rightarrow 2H_2O(l) + $ **$CaCl_2$**(aq)

72.   a.   soluble (Rule 2: Most potassium salts are soluble.)

b.   soluble (Rule 2: Most ammonium salts are soluble.)

c.   insoluble (Rule 6: Most carbonate salts are only slightly soluble.)

d.   insoluble (Rule 6: Most phosphate salts are only slightly soluble.)

e.   soluble (Rule 2: Most sodium salts are soluble.)

f.   insoluble (Rule 6: Most carbonate salts are only slightly soluble.)

g.   soluble (Rule 3: Most chloride salts are soluble.)

74.   The precipitates are marked in boldface type.

a.   Rule 3: AgCl is listed as an exception

$AgNO_3(aq) + HCl(aq) \rightarrow$ **AgCl**$(s) + HNO_3(aq)$

b.   Rule 6: Most cabonate salts are only slightly soluble.

$CuSO_4(aq) + (NH_4)_2CO_3(aq) \rightarrow$ **CuCO$_3$**$(s) + (NH_4)_2SO_4(aq)$

c.   Rule 6: Most carbonate salts are only slightly soluble.

$FeSO_4(aq) + K_2CO_3(aq) \rightarrow$ **FeCO$_3$**$(s) + K_2SO_4(aq)$

d.   no reaction

e.   Rule 6: Most carbonate salts are only slightly soluble.

$Pb(NO_3)_2(aq) + Li_2CO_3(aq) \rightarrow$ **PbCO$_3$**$(s) + 2LiNO_3(aq)$

f.   Rule 5: Most hydroxide compounds are only slightly soluble.

$SnCl_4(aq) + 4NaOH(aq) \rightarrow$ **Sn(OH)$_4$**$(s) + 4NaCl(aq)$

76. $Fe^{2+}(aq) + S^{2-}(aq) \rightarrow FeS(s)$

$2Cr^{3+}(aq) + 3S^{2-}(aq) \rightarrow Cr_2S_3(s)$

$Ni^{2+}(aq) + S^{2-}(aq) \rightarrow NiS(s)$

78.   These anions tend to form insoluble precipitates with *many* metal ions. The following are illustrative for cobalt(II) chloride, tin(II) chloride, and copper(II) nitrate reacting with the sodium salts of the given anions.

a. $CoCl_2(aq) + Na_2S(aq) \rightarrow CoS(s) + 2NaCl(aq)$

$SnCl_2(aq) + Na_2S(aq) \rightarrow SnS(s) + 2NaCl(aq)$

$Cu(NO_3)_2(aq) + Na_2S(aq) \rightarrow CuS(s) + 2NaNO_3(aq)$

b. $CoCl_2(aq) + Na_2CO_3(aq) \rightarrow CoCO_3(s) + 2NaCl(aq)$

$SnCl_2(aq) + Na_2CO_3(aq) \rightarrow SnCO_3(s) + 2NaCl(aq)$

$Cu(NO_3)_2(aq) + Na_2CO_3(aq) \rightarrow CuCO_3(s) + 2NaNO_3(aq)$

c. $CoCl_2(aq) + 2NaOH(aq) \rightarrow Co(OH)_2(s) + 2NaCl(aq)$

$SnCl_2(aq) + 2NaOH(aq) \rightarrow Sn(OH)_2(s) + 2NaCl(aq)$

$Cu(NO_3)_2(aq) + 2NaOH(aq) \rightarrow Cu(OH)_2(s) + 2NaNO_3(aq)$

d. $3CoCl_2(aq) + 2Na_3PO_4(aq) \rightarrow Co_3(PO_4)_2(s) + 6NaCl(aq)$

$3SnCl_2(aq) + 2Na_3PO_4(aq) \rightarrow Sn_3(PO_4)_2(s) + 6NaCl(aq)$

$3Cu(NO_3)_2(aq) + 2Na_3PO_4(aq) \rightarrow Cu_3(PO_4)_2(s) + 6NaNO_3(aq)$

80. a. $Na + O_2 \rightarrow Na_2O_2$

Balance sodium: $\mathbf{2}Na + O_2 \rightarrow Na_2O_2$

Balanced equation: $2Na(s) + O_2(g) \rightarrow Na_2O_2(s)$

b. $Fe(s) + H_2SO_4(aq) \rightarrow FeSO_4(aq) + H_2(g)$

Equation is already balanced!

c. $Al_2O_3 \rightarrow Al + O_2$

Balance oxygen: $\mathbf{2}Al_2O_3 \rightarrow Al + 3O_2$

Balance aluminum: $2Al_2O_3 \rightarrow \mathbf{4}Al + 3O_2$

Balanced equation: $2Al_2O_3(s) \rightarrow 4Al(s) + 3O_2(g)$

d. $Fe + Br_2 \rightarrow FeBr_3$

Balance bromine: $Fe + \mathbf{3}Br_2 \rightarrow \mathbf{2}FeBr_3$

Balance iron: $\mathbf{2}Fe + 3Br_2 \rightarrow 2FeBr_3$

Balanced equation: $2Fe(s) + 3Br_2(l) \rightarrow 2FeBr_3(s)$

e. $Zn + HNO_3 \rightarrow Zn(NO_3)_2 + H_2$

Balance nitrate ions: $Zn + \mathbf{2}HNO_3 \rightarrow Zn(NO_3)_2 + H_2$

Balanced equation: $Zn(s) + 2HNO_3(aq) \rightarrow$

$Zn(NO_3)_2(aq) + H_2(g)$

82. a. $2C_4H_{10}(l) + 13O_2(g) \rightarrow 8CO_2(g) + 10H_2O(g)$

b. $C_4H_{10}O(l) + 6O_2(g) \rightarrow 4CO_2(g) + 5H_2O(g)$

c. $2C_4H_{10}O_2(l) + 11O_2(g) \rightarrow 8CO_2(g) + 10H_2O(g)$

84. a. $2NaHCO_3(s) \rightarrow Na_2CO_3(s) + H_2O(g) + CO_2(g)$

b. $2NaClO_3(s) \rightarrow 2NaCl(s) + 3O_2(g)$

c. $2HgO(s) \rightarrow 2Hg(l) + O_2(g)$

d. $C_{12}H_{22}O_{11}(s) \rightarrow 12C(s) + 11H_2O(g)$

e. $2H_2O_2(l) \rightarrow 2H_2O(l) + O_2(g)$

86. $Fe(s) + H_2SO_4(aq) \rightarrow FeSO_4(aq) + H_2(g)$

$Zn(s) + H_2SO_4(aq) \rightarrow ZnSO_4(aq) + H_2(g)$

$Mg(s) + H_2SO_4(aq) \rightarrow MgSO_4(aq) + H_2(g)$

$Co(s) + H_2SO_4(aq) \rightarrow CoSO_4(aq) + H_2(g)$

$Ni(s) + H_2SO_4(aq) \rightarrow NiSO_4(aq) + H_2(g)$

88.   a.   one

      b.   one

      c.   two

      d.   two

      e.   three

90.   A very simple example that fits the bill is: $C(s) + O_2(g) \rightarrow CO_2(g)$

92.   a. $2C_3H_8O(l) + 9O_2(g) \rightarrow 6CO_2(g) + 8H_2O(g)$

           oxidation-reduction, combustion

      b. $HCl(aq) + AgC_2H_3O_2(aq) \rightarrow AgCl(s) + HC_2H_3O_2(aq)$

           precipitation, double-displacement

      c. $3HCl(aq) + Al(OH)_3(s) \rightarrow AlCl_3(aq) + 3H_2O(l)$

           acid-base, double-displacement

      d. $2H_2O_2(aq) \rightarrow 2H_2O(l) + O_2(g)$

           oxidation-reduction, decomposition

      e. $N_2H_4(l) + O_2(g) \rightarrow N_2(g) + 2H_2O(g)$

           oxidation-reduction, combustion

94. $2Na(s) + Cl_2(g) \rightarrow 2NaCl(s)$

      $2Al(s) + 3Cl_2(g) \rightarrow 2AlCl_3(s)$

      $Zn(s) + Cl_2(g) \rightarrow ZnCl_2(s)$

      $Ca(s) + Cl_2(g) \rightarrow CaCl_2(s)$

      $2Fe(s) + 3Cl_2(g) \rightarrow 2FeCl_3(s)$; $Fe(s) + Cl_2(g) \rightarrow FeCl_2(s)$

Cumulative Review: Chapters 6 and 7

2.   A chemical equation indicates the substances necessary for a chemical reaction to take place, as well as what is produced by that chemical reaction. The substances to the left of the arrow in a chemical equation are called the reactants; those to the right of the arrow are referred to as the products. In addition, if a chemical equation has been balanced, then the equation indicates the relative proportions in which the reactant molecules combine to form the product molecules.

4.   It is *never* permissible to change the subscripts of a formula when balancing a chemical equation: changing the subscripts changes the *identity* of a substance from one chemical to another. For example, consider the unbalanced chemical equation

$$H_2(g) + O_2(g) \rightarrow H_2O(l).$$

If you changed the *formula* of the product from $H_2O(l)$ to $H_2O_2(l)$, the equation would appear to be "balanced". However, H_2O is water, whereas H_2O_2 is hydrogen peroxide-a completely different chemical substance (which is not prepared by reaction of the elements hydrogen and oxygen).

When we balance a chemical equation, it is permitted only to adjust the *coefficients* of a formula, so changing a coefficient merely changes the number of molecules of a substance being used in the reaction, without changing the identity of the substance. For the example above, we can balance the equation by putting coefficients of 2 in front of the formulas of H_2 and H_2O: these coefficients do not change the nature of what is reacting and what product is formed.

$$2H_2(g) + O_2(g) \rightarrow 2H_2O(l)$$

6.   A precipitation reaction is one in which a *solid* forms when the reactants are combined: the solid is called a precipitate. If you were to perform such a reaction, the mixture would turn cloudy as the reactants are combined, and a solid would eventually settle from the mixture on standing. There are many examples of such precipitation reactions: consult the solubility rules in Table 7.1 if you need help. One example would be to combine barium nitrate and sodium carbonate solutions: a precipitate of barium carbonate would form.

$$Ba(NO_3)_2(aq) + Na_2CO_3(aq) \rightarrow BaCO_3(s) + 2NaNO_3(aq)$$

8.   In summary, nearly all compounds containing the nitrate, sodium, potassium, and ammonium ions are soluble in water. Most salts containing the chloride and sulfate ions are soluble in water, with specific exceptions (see Table 7.1 for these exceptions). Most compounds containing the hydroxide, sulfide, carbonate, and phosphate ions are not soluble in water, unless the compound also contains one of the cations mentioned above (Na^+, K^+, NH_4^+).

The solubility rules are phrased as if you had a sample of a given solute and wanted to see if you could dissolve it in water. These rules can also be applied, however, to predict the identity of the solid produced in a precipitation reaction: a given combination of ions will not be soluble in water whether you take a pure compound out of a reagent bottle or if you generate the insoluble combination of ions during a chemical reaction. For example, the solubility rules say that $BaSO_4$ is not soluble in water. This means not only that a pure sample of $BaSO_4$ taken from a reagent bottle will not dissolve in water, but also that if Ba^{2+} ion and SO_4^{2-} ion end up together in the same solution, they will precipitate as $BaSO_4$. If we were to combine barium chloride and sulfuric acid solutions

$$BaCl_2(aq) + H_2SO_4(aq) \rightarrow BaSO_4(s) + 2HCl(aq)$$

then, because barium sulfate is not soluble in water, a precipitate of $BaSO_4(s)$ would form. Because a precipitate of $BaSO_4(s)$ would form no matter what barium compound or what sulfate compound were mixed, we can write the net ionic equation for the reaction as

$$Ba^{2+}(aq) + SO_4^{2-}(aq) \rightarrow BaSO_4(s).$$

Thus if, for example, barium nitrate solution were combined with sodium sulfate solution, a precipitate of $BaSO_4$ would form. Barium sulfate is insoluble in water regardless of its source.

10.   Acids (such as the citric acid found in citrus fruits and the acetic acid found in vinegar) were first noted primarily because of their sour taste. The first bases noted were characterized by their bitter taste and slippery feel on the skin. Acids and bases chemically react with (neutralize) each other forming water: the net ionic equation is

$$H^+(aq) + OH^-(aq) \rightarrow H_2O(l).$$

The *strong* acids and bases fully ionize when they dissolve in water: because these substances fully ionize, they are strong electrolytes. The common strong acids are HCl(hydrochloric), HNO_3(nitric), H_2SO_4(sulfuric), and $HClO_4$(perchloric). The most common strong bases are the alkali metal hydroxides, particularly NaOH(sodium hydroxide) and KOH(potassium hydroxide).

12.   Oxidation-reduction reactions are electron-transfer reactions. Oxidation represents a loss of electrons by an atom, molecule, or ion, whereas reduction is the gain of electrons by such a species. Because an oxidation-reduction process represents the transfer of electrons between species, you can't have one without the other also taking place: the electrons lost by one species must be gained by some other species. An example of a simple oxidation reduction reaction between a metal and a nonmetal could be the following

$$Mg(s) + F_2(g) \rightarrow MgF_2(s).$$

In this process, Mg atoms lose two electrons each to become Mg^{2+} ions in MgF_2: Mg is oxidized. Each F atom of F_2 gains one electron to become an

F^- ion, for a total of two electrons gained for each F_2 molecule: F_2 is reduced.

$$Mg \rightarrow Mg^{2+} + 2e^- \qquad\qquad 2(F + e^- \rightarrow F^-)$$

14.   In general, a synthesis reaction represents the reaction of elements or simple compounds to produce more complex substances. There are many examples of synthesis reactions, for example

$$N_2(g) + 3H_2(g) \rightarrow 2NH_3(g)$$

$$NaOH(aq) + CO_2(g) \rightarrow NaHCO_3(s).$$

Decomposition reactions represent the breakdown of a more complex substance into simpler substances. There are many examples of decomposition reactions, for example

$$2H_2O_2(aq) \rightarrow 2H_2O(l) + O_2(g).$$

Synthesis and decomposition reactions are very often also oxidation-reduction reactions, especially if an elemental substance reacts or is generated. It is not necessary, however, for synthesis and decomposition reactions to always involve oxidation-reduction. The reaction between NaOH and CO_2 given as an example of a synthesis reaction does *not* represent oxidation-reduction.

16.   a. $2Na(s) + 2H_2O(l) \rightarrow 2NaOH(aq) + H_2(g)$

$2K(s) + 2H_2O(l) \rightarrow 2KOH(aq) + H_2(g)$

b. $2Na(s) + Cl_2(g) \rightarrow 2NaCl(s)$

$2K(s) + Cl_2(g) \rightarrow 2KCl(s)$

c. $3Na(s) + P(s) \rightarrow Na_3P(s)$

$3K(s) + P(s) \rightarrow K_3P(s)$

d. $6Na(s) + N_2(g) \rightarrow 2Na_3N(s)$

$6K(s) + N_2(g) \rightarrow 2K_3N(s)$

e. $2Na(s) + H_2(g) \rightarrow 2NaH(s)$

$2K(s) + H_2(g) \rightarrow 2KH(s)$

18.   a. $Ba(NO_3)_2(aq) + K_2CrO_4(aq) \rightarrow BaCrO_4(s) + 2KNO_3(aq)$

b. $NaOH(aq) + CH_3COOH(aq) \rightarrow H_2O(l) + NaCH_3COO(aq)$
(then evaporate the water from the solution)

c. $AgNO_3(aq) + NaCl(aq) \rightarrow AgCl(s) + NaNO_3(aq)$

d. $Pb(NO_3)_2(aq) + H_2SO_4(aq) \rightarrow PbSO_4(s) + 2HNO_3(aq)$

e. $2NaOH(aq) + H_2SO_4(aq) \rightarrow Na_2SO_4(aq) + 2H_2O(l)$
(then evaporate the water from the solution)

f. $Ba(NO_3)_2(aq) + Na_2CO_3(aq) \rightarrow BaCO_3(s) + 2NaNO_3(aq)$

20.   a. $FeO(s) + 2HNO_3(aq) \rightarrow Fe(NO_3)_2(aq) + H_2O(l)$
acid-base, double-displacement

b. $2Mg(s) + 2CO_2(g) + O_2(g) \rightarrow 2MgCO_3(s)$
synthesis; oxidation-reduction

c. $2NaOH(s) + CuSO_4(aq) \rightarrow Cu(OH)_2(s) + Na_2SO_4(aq)$
precipitation, double-displacement

d. $HI(aq) + KOH(aq) \rightarrow KI(aq) + H_2O(l)$
acid-base, double-displacement

e. $C_3H_8(g) + 5O_2(g) \rightarrow 3CO_2(g) + 4H_2O(g)$
combustion; oxidation-reduction

f. $Co(NH_3)_6Cl_2(s) \rightarrow CoCl_2(s) + 6NH_3(g)$
decomposition

g. $2HCl(aq) + Pb(C_2H_3O_2)_2(aq) \rightarrow 2HC_2H_3O_2(aq) + PbCl_2(s)$
precipitation, double-displacement

h. $C_{12}H_{22}O_{11}(s) \rightarrow 12C(s) + 11H_2O(g)$
decomposition; oxidation-reduction

i. $2Al(s) + 6HNO_3(aq) \rightarrow 2Al(NO_3)_3(aq) + 3H_2(g)$
oxidation-reduction; single-displacement

j. $4B(s) + 3O_2(g) \rightarrow 2B_2O_3(s)$
synthesis; oxidation-reduction

22.   Specific examples will depend on students' responses. The following are general equations that illustrate each type of reaction:

precipitation: typical when solutions of two ionic solutes are mixed, and one of the new combinations of ions is insoluble.

$$A^+B^-(aq) + C^+D^-(aq) \rightarrow AD(s) + C^+D^-(aq)$$

single displacement: one element replaces a less reactive element from a compound.

$$A(s) + B^+C^-(aq) \rightarrow A^+C^-(aq) + B(s)$$

combustion: a rapid oxidation reaction, most commonly involving $O_2(g)$. Most examples in the text involve the combustion of hydrocarbons or hydrocarbon derivatives.

$$(\text{hydrocarbon or derivative}) + O_2(g) \rightarrow CO_2(g) + H_2O(g)$$

synthesis: elements or simple compounds combine to make more complicated molecules.

$$A(s) + B(s) \rightarrow AB(s)$$

oxidation-reduction: reactions in which electrons are transferred from one species to another. Examples of oxidation-reduction reactions include single-displacement, combustion, synthesis, and decomposition reactions.

decomposition: a compound breaks down into elements and/or simpler compounds.

$$AB(s) \rightarrow A(s) + B(s)$$

acid-base neutralization: a neutralization takes place when a proton from an acid combines with a hydroxide ion from a base to make a water molecule.

Chapter 8   Chemical Composition

2. $500. \text{ g} \times \dfrac{1 \text{ cork}}{1.63 \text{ g}} = 306.7 = 307 \text{ corks}$

$500. \text{ g} \times \dfrac{1 \text{ stopper}}{4.31 \text{ g}} = 116 \text{ stoppers}$

1 kg (1000 g) of corks contains $\left(1000 \text{ g} \times \dfrac{1 \text{ cork}}{1.63 \text{ g}}\right) = 613.49 = 613 \text{ corks}$

613 stoppers would weigh $\left(613 \text{ stoppers} \times \dfrac{4.31 \text{ g}}{1 \text{ stopper}}\right) = 2644 \text{ g} = 2640 \text{ g}$

The ratio of the mass of a stopper to the mass of a cork is (4.31 g/1.63 g). So the mass of stoppers that contains the same number of stoppers as there are corks in 1000 g of corks is

$1000 \text{ g} \times \dfrac{4.31 \text{ g}}{1.63 \text{ g}} = 2644 \text{ g} = 2640 \text{ g}.$

4.   We use the average atomic mass of an element when performing calculations because the average mass takes into account the individual masses and relative abundances of all the isotopes of an element.

6.   a. $52.00 \text{ amu} \times \dfrac{1 \text{ Cr atom}}{52.00 \text{ amu}} = 1.000 \text{ Cr atom} = 1 \text{ Cr atom}$

b. $749.2 \text{ amu} \times \dfrac{1 \text{ As atom}}{74.92 \text{ amu}} = 10.00 \text{ As atoms} = 10 \text{ As atoms}$

c. $4274 \text{ amu} \times \dfrac{1 \text{ Rb atom}}{85.47 \text{ amu}} = 50.00 \text{ Rb atoms} = 50 \text{ Rb atoms}$

d. $2698 \text{ amu} \times \dfrac{1 \text{ Al atom}}{26.98 \text{ amu}} = 100.0 \text{ Al atoms} = 100 \text{ Al atoms}$

e. $1900. \text{ amu} \times \dfrac{1 \text{ F atom}}{19.00 \text{ amu}} = 100.0 \text{ F atoms} = 100 \text{ F atoms}$

8.   From the table of elements on the inside cover of this book, an average iodine atom has a mass of 126.9 amu. So the number of iodine atoms contained in the given sample would be

$7.043 \times 10^4 \text{ amu} \times \dfrac{1 \text{ I atom}}{126.9 \text{ amu}} = 555 \text{ I atoms}$

and the mass of 451 iodine atoms would be given by

$451 \text{ I atoms} \times \dfrac{126.9 \text{ amu}}{1 \text{ I atom}} = 5.72 \times 10^4 \text{ amu}.$

10.   twice Avogadro's number $(2 \times 6.022 \times 10^{23} = 1.201 \times 10^{24} \text{ atoms}; \ 2.00 \text{ mol})$

12. The ratio of the atomic mass of Fe to the atomic mass of N is (55.85 amu/14.01 amu), and the mass of iron is given by

$$14.01 \text{ g} \times \frac{55.85 \text{ amu}}{14.01 \text{ amu}} = 55.85 \text{ g Fe.}$$

14. The ratio of the atomic mass of Co to the atomic mass of F is (58.93 amu/19.00 amu), and the mass of cobalt is given by

$$57.0 \text{ g} \times \frac{58.93 \text{ amu}}{19.00 \text{ amu}} = 177 \text{ g Co.}$$

16. 1 mol O = 16.00 g O = 6.02×10^{23} O atoms

$$1 \text{ O atom} \times \frac{16.00 \text{ g O}}{6.022 \times 10^{23} \text{ O atoms}} = 2.66 \times 10^{-23} \text{ g}$$

18.
$$0.50 \text{ mol O atoms} \times \frac{16.00 \text{ g}}{1 \text{ mol}} = 8.0 \text{ g O}$$

$$4 \text{ mol H atoms} \times \frac{1.008 \text{ g}}{1 \text{ mol}} = 4 \text{ g H}$$

Half a mole of O atoms weighs more than 4 moles of H atoms.

20. a. $1.71 \times 10^{-3} \text{ g Ag} \times \dfrac{1 \text{ mol Ag}}{107.9 \text{ g}} = 1.58 \times 10^{-5} \text{ mol Ag}$

    b. $280.9 \text{ mg Si} \times \dfrac{1 \text{ g}}{1000 \text{ mg}} \times \dfrac{1 \text{ mol Si}}{28.09 \text{ g}} = 1.000 \times 10^{-2} \text{ mol Si}$

    c. $63.45 \text{ g Zn} \times \dfrac{1 \text{ mol Zn}}{65.38 \text{ g}} = 0.9705 \text{ mol Zn}$

    d. $5.869 \text{ g Ni} \times \dfrac{1 \text{ mol Ni}}{58.69 \text{ g}} = 0.1000 \text{ mol Ni}$

    e. $128.6 \text{ g U} \times \dfrac{1 \text{ mol U}}{238.0 \text{ g}} = 0.5403 \text{ mol U}$

    f. $3.251 \text{ kg Li} \times \dfrac{1000 \text{ g}}{1 \text{ kg}} \times \dfrac{1 \text{ mol Li}}{6.941 \text{ g}} = 4.684 \times 10^{2} \text{ mol Li}$

    g. $12.01 \text{ g C} \times \dfrac{1 \text{ mol C}}{12.01 \text{ g}} = 1.000 \text{ mol C}$

22. a. $1.76 \times 10^{-3} \text{ mol Cs} \times \dfrac{132.9 \text{ g}}{1 \text{ mol}} = 2.34 \times 10^{-1} \text{ g Cs}$

b. $0.0125 \text{ mol Ne} \times \dfrac{20.18 \text{ g}}{1 \text{ mol}} = 0.252 \text{ g Ne}$

c. $5.29 \times 10^3 \text{ mol Pb} \times \dfrac{207.2 \text{ g}}{1 \text{ mol}} = 1.10 \times 10^6 \text{ g Pb}$

d. $0.00000122 \text{ mol Na} \times \dfrac{22.99 \text{ g}}{1 \text{ mol}} = 2.80 \times 10^{-5} \text{ g Na}$

e. $5.51 \text{ millimol As} \times \dfrac{74.92 \text{ mg}}{1 \text{ millimol}} = 413 \text{ mg} = 0.413 \text{ g As}$

f. $8.72 \text{ mol C} \times \dfrac{12.01 \text{ g}}{1 \text{ mol}} = 105 \text{ g C}$

24.   a. $0.00103 \text{ g Co} \times \dfrac{6.022 \times 10^{23} \text{ Co atoms}}{58.93 \text{ g Co}} = 1.05 \times 10^{19} \text{ Co atoms}$

b. $0.00103 \text{ mol Co} \times \dfrac{6.022 \times 10^{23} \text{ Co atoms}}{1 \text{ mol}} = 6.20 \times 10^{20} \text{ Co atoms}$

c. $2.75 \text{ g cobalt} \times \dfrac{1 \text{ mol}}{58.93 \text{ g Co}} = 0.0467 \text{ mol Co}$

d. $5.99 \times 10^{21} \text{ Co atoms} \times \dfrac{1 \text{ mol}}{6.022 \times 10^{23} \text{ Co atoms}} = 0.00995 \text{ mol Co}$

e. $4.23 \text{ mol Co} \times \dfrac{58.93 \text{ g Co}}{1 \text{ mol Co}} = 249 \text{ g Co}$

f. $4.23 \text{ mol Co} \times \dfrac{6.022 \times 10^{23} \text{ Co atoms}}{1 \text{ mol Co}} = 2.55 \times 10^{24} \text{ Co atoms}$

g. $4.23 \text{ g Co} \times \dfrac{6.022 \times 10^{23} \text{ Co atoms}}{58.93 \text{ g Co}} = 4.32 \times 10^{22} \text{ Co atoms}$

26.   adding together (summing)

28.   a. NO_2 nitrogen dioxide

mass of 1 mol N =	14.01 g =	14.01 g
mass of 2 mol) =	2(16.00 g) =	32.00 g
molar mass of NO_2 =		46.01 g

b. N_2O dinitrogen monoxide

mass of 2 mol N =	2(14.01) g =	28.02 g
mass of 1 mol O =	16.00 g =	16.00 g
molar mass of N_2O =		44.02 g

c. XeF_4 xenon tetrafluoride

mass of 1 mol Xe =   131.3 g =   131.3 g

mass of 4 mol F =   4(19.00 g) =   76.00 g

molar mass of XeF_4 =   207.3 g

d.   NaOCl sodium hypochlorite

mass of 1 mol Na =   22.99 g =   22.99 g

mass of 1 mol O=   16.00 g =   16.00 g

mass of 1 mol Cl =   35.45 g =   35.45 g

molar mass of NaOCl =   74.44 g

e. HNO_3 nitric acid

mass of 1 mol H =   1.008 g =   1.008 g

mass of 1 mol N =   14.01 g =   14.01 g

mass of 3 mol O =   3(16.00 g) =   48.00 g

molar mass of HNO_3 =   63.018 g = 63.02 g

f. $NaC_2H_3O_2$ sodium acetate

mass of 1 mol Na =   22.99 g =   22.99 g

mass of 2 mol C =   2(12.01 g) =   24.02 g

mass of 3 mol H =   3(1.008 g) =   3.024 g

mass of 2 mol O =   2(16.00 g) =   32.00 g

molar mass of H_2SO_4 =   82.034 g = 82.03 g

30.   a.   mass of 2 mol N =   2(14.01 g) =   28.02 g

mass of 8 mol H =   8(1.008 g) =   8.064 g

mass of 1 mol S =   32.07 g =   32.07 g

molar mass of $(NH_4)_2S$ =   68.154 g = 68.15 g

b.   mass of 6 mol C =   6(12.01 g) =   72.06 g

mass of 4 mol H =   4(1.008 g) =   4.032 g

mass of 1 mol O =   16.00 g =   16.00 g

mass of 2 mol Cl =   2(35.45 g) =   70.90 g

molar mass of $C_6H_4OCl_2$ =   162.992 g = 162.99 g

c.   mass of 1 mol Ba =   137.33 g =   137.33 g

mass of 2 mol H =   2(1.008 g) = 2.016 g

molar mass of BaH_2 =   139.346 g = 139.35 g

d.   mass of 1 mol K =   39.10 g =   39.10 g

mass of 2 mol H =   2(1.008 g) =   2.016 g

mass of 1 mol P =   30.97 g =   30.97 g

mass of 4 mol O =   4(16.00 g) =   64.00 g

molar mass of KH_2PO_4 =   136.086 g = 136.09 g

e.   mass of 2 mol K =   2(39.10 g) =   78.20 g

mass of 1 mol H =   1.008 g =   1.008 g

mass of 1 mol P =   30.97 g =   30.97 g

mass of 4 mol O =   4(16.00 g) =   64.00 g

molar mass of K_2HPO_4 =   174.178 g = 174.18 g

f.   mass of 3 mol K =   3(39.10 g) = 117.3 g

mass of 1 mol P =   30.97 g =   30.97 g

mass of 4 mol O =   4(16.00 g) =   64.00 g

molar mass of K_3PO_4 =   212.27 g = 212.3 g

32. a.   molar mass of HBr = 80.91 g

$$92.4 \text{ g HBr} \times \frac{1 \text{ mol}}{80.91 \text{ g}} = 1.14 \text{ mol}$$

b.   molar mass of $FeCl_3$ = 162.2 g;   5.34 mg = 0.00534 g

$$0.00534 \text{ g FeCl}_3 \times \frac{1 \text{ mol}}{162.2 \text{ g}} = 3.29 \times 10^{-5} \text{ mol FeCl}_3$$

c.   molar mass of H_2SO_4 = 98.09 g;   2.21 kg = 2210 g

$$2210 \text{ g H}_2\text{SO}_4 \times \frac{1 \text{ mol}}{98.09 \text{ g}} = 22.5 \text{ mol H}_2\text{SO}_4$$

d.   molar mass of $BaCO_3$ = 197.3 g

$$3.44 \text{ g} \times \frac{1 \text{ mol}}{197.3 \text{ g}} = 0.0174 \text{ mol BaCO}_3$$

e.   molar mass of $AlCl_3$ = 133.3 g

$$2.89 \text{ g AlCl}_3 \times \frac{1 \text{ mol}}{133.3 \text{ g}} = 0.0217 \text{ mol AlCl}_3$$

f.   molar mass of Li_2CO_3 = 73.89 g

$$7.21 \text{ g Li}_2\text{CO}_3 \times \frac{1 \text{ mol}}{73.89 \text{ g}} = 0.0976 \text{ mol Li}_2\text{CO}_3$$

34. a.   molar mass of NaH_2PO_4 = 120.0 g

$$4.26 \times 10^{-3} \text{ g } NaH_2PO_4 \times \frac{1 \text{ mol}}{120.0 \text{ g}} = 3.55 \times 10^{-5} \text{ mol } NaH_2PO_4$$

b.   molar mass of CuCl = 99.00 g

$$521 \text{ g CuCl} \times \frac{1 \text{ mol}}{99.00 \text{ g}} = 5.26 \text{ mol CuCl}$$

c.   molar mass of Fe = 55.85 g

$$151 \text{ kg Fe} \times \frac{1000 \text{ g}}{1 \text{ kg}} \times \frac{1 \text{ mol}}{55.85 \text{ g}} = 2.70 \times 10^3 \text{ mol Fe}$$

d.   molar mass of SrF_2 = 125.6 g

$$8.76 \text{ g } SrF_2 \times \frac{1 \text{ mol}}{125.6 \text{ g}} = 0.0697 \text{ mol } SrF_2$$

e.   molar mass of Al = 26.98 g

$$1.26 \times 10^4 \text{ g Al} \times \frac{1 \text{ mol}}{26.98 \text{ g}} = 467 \text{ mol Al}$$

36. a.   molar mass of CO = 28.01 g

$$0.00471 \text{ mol} \times \frac{28.01 \text{ g}}{1 \text{ mol}} = 0.132 \text{ g}$$

b.   molar mass of $AuCl_3$ = 303.4 g

$$1.75 \times 10^{-6} \text{ mol } AuCl_3 \times \frac{303.4 \text{ g}}{1 \text{ mol}} = 5.31 \times 10^{-4} \text{ g}$$

c.   molar mass of $FeCl_3$ = 162.2 g

$$228 \text{ mol } FeCl_3 \times \frac{162.2 \text{ g}}{1 \text{ mol}} = 3.70 \times 10^4 \text{ g}$$

d.   molar mass of K_3PO_4 = 212.3 g;   2.98 millimol = 0.00298 mol

$$0.00298 \text{ mol } K_3PO_4 \times \frac{212.3 \text{ g}}{1 \text{ mol}} = 0.633 \text{ g}$$

e.   molar mass of LiCl = 42.39 g

$$2.71 \times 10^{-3} \text{ mol LiCl} \times \frac{42.39 \text{ g}}{1 \text{ mol}} = 0.115 \text{ g}$$

f.   molar mass of NH_3 = 17.03 g

$$6.55 \text{ mol } NH_3 \times \frac{17.03 \text{ g}}{1 \text{ mol}} = 112 \text{ g}$$

38. a.   molar mass of CO_2 = 44.01 g

$$1.27 \text{ mmol} \times \frac{1 \text{ mol}}{10^3 \text{ mmol}} \times \frac{44.01 \text{ g}}{1 \text{ mol}} = 0.0559 \text{ g } CO_2$$

b.   molar mass of NCl_3 = 120.4 g

$$4.12 \times 10^3 \text{ mol } NCl_3 \times \frac{120.4 \text{ g}}{1 \text{ mol}} = 4.96 \times 10^5 \text{ g } NCl_3$$

c.   molar mass of NH_4NO_3 = 80.05 g

$$0.00451 \text{ mol } NH_4NO_3 \times \frac{80.05 \text{ g}}{1 \text{ mol}} = 0.361 \text{ g } NH_4NO_3$$

d.   molar mass of H_2O = 18.02 g

$$18.0 \text{ mol } H_2O \times \frac{18.02 \text{ g}}{1 \text{ mol}} = 324 \text{ g } H_2O$$

e.   molar mass of $CuSO_4$ = 159.6 g

$$62.7 \text{ mol } CuSO_4 \times \frac{159.6 \text{ g}}{1 \text{ mol}} = 1.00 \times 10^4 \text{ g } CuSO_4$$

40. a. $6.37 \text{ mol CO} \times \dfrac{6.022 \times 10^{23} \text{ molecules}}{1 \text{ mol}} = 3.84 \times 10^{24} \text{ molecules CO}$

b.   molar mass of CO = 28.01 g

$$6.37 \text{ g} \times \frac{1 \text{ mol}}{28.01 \text{ g}} \times \frac{6.022 \times 10^{23} \text{ molec.}}{1 \text{ mol}} = 1.37 \times 10^{23} \text{ molec. CO}$$

c.   molar mass of H_2O = 18.02 g

$$2.62 \times 10^{-6} \text{ g} \times \frac{6.022 \times 10^{23} \text{ molec.}}{18.02 \text{ g}} = 8.76 \times 10^{16} \text{ molec. } H_2O$$

d. $2.62 \times 10^{-6} \text{ g} \times \dfrac{6.022 \times 10^{23} \text{ molec.}}{1 \text{ mol}} = 1.58 \times 10^{18} \text{ molec. } H_2O$

e.   molar mass of C_6H_6 = 78.11 g

$$5.23 \text{ g} \times \frac{6.022 \times 10^{23} \text{ molecules}}{78.11 \text{ g}} = 4.03 \times 10^{22} \text{ molecules } C_6H_6$$

42. a.   molar mass of Na_2SO_4 = 142.1 g

$$2.01 \text{ g Na}_2SO_4 \times \frac{1 \text{ mol Na}_2SO_4}{142.1 \text{ g}} \times \frac{1 \text{ mol S}}{1 \text{ mol Na}_2SO_4} = 0.0141 \text{ mol S}$$

b.   molar mass of Na_2SO_3 = 126.1 g

$$2.01 \text{ g Na}_2SO_3 \times \frac{1 \text{ mol Na}_2SO_3}{126.1 \text{ g}} \times \frac{1 \text{ mol S}}{1 \text{ mol Na}_2SO_3} = 0.0159 \text{ mol S}$$

c.   molar mass of Na_2S = 78.05 g

$$2.01 \text{ g Na}_2S \times \frac{1 \text{ mol Na}_2S}{78.05 \text{ g}} \times \frac{1 \text{ mol S}}{1 \text{ mol Na}_2S} = 0.0258 \text{ mol S}$$

d.   molar mass of $Na_2S_2O_3$ = 158.1 g

$$2.01 \text{ g Na}_2S_2O_3 \times \frac{1 \text{ mol Na}_2S_2O_3}{158.1 \text{ g}} \times \frac{2 \text{ mol S}}{1 \text{ mol Na}_2S_2O_3} = 0.0254 \text{ mol S}$$

44.   less than

46. a.   mass of Na present =   2(22.99 g) =   45.98 g

mass of S present =   32.07 g =   32.07 g

molar mass of Na_2SO_4 =   78.05 g

$$\%Na = \frac{45.98 \text{ g Na}}{78.05 \text{ g}} \times 100 = 58.91\% \text{ Na}$$

$$\%S = \frac{32.07 \text{ g S}}{78.05 \text{ g}} \times 100 = 41.09\% \text{ S}$$

b.   mass of N present = 2(14.01 g) =   28.02 g

mass of H present =   4(1.008 g) = 4.032 g

mass of O present = 2(16.00 g) =   32.00 g

molar mass of NH_4NO_2 =   64.05 g

$$\%N = \frac{28.02 \text{ g N}}{64.05 \text{ g}} \times 100 = 43.75\% \text{ N}$$

$$\%H = \frac{4.032 \text{ g H}}{64.05 \text{ g}} \times 100 = 6.295\% \text{ H}$$

$$\%O = \frac{32.00 \text{ g O}}{64.05 \text{ g}} \times 100 = 49.96\% \text{ O}$$

c.   mass of N present = 2(14.01 g) =   28.02 g

mass of H present =   4(1.008 g) = 4.032 g

mass of O present = 3(16.00) g =   48.00 g

molar mass of NH_4NO_3 =   80.05 g

$$\%N = \frac{28.02 \text{ g N}}{80.05 \text{ g}} \times 100 = 35.00\% \text{ N}$$

$$\%H = \frac{4.032 \text{ g H}}{80.05 \text{ g}} \times 100 = 5.037\% \text{ H}$$

$$\%O = \frac{48.00 \text{ g O}}{80.05 \text{ g}} \times 100 = 59.96\% \text{ O}$$

d.   mass of N present =   14.01 g =   14.01 g

mass of H present = 2(1.008 g) = 2.016 g

mass of Cl present = 35.45 g =   35.45 g

molar mass of NH_2Cl =   51.48 g

$$\%N = \frac{14.01 \text{ g N}}{51.48 \text{ g}} \times 100 = 27.22\% \text{ N}$$

$$\%H = \frac{2.016 \text{ g H}}{51.48 \text{ g}} \times 100 = 3.916\% \text{ H}$$

$$\%Cl = \frac{35.45 \text{ g Cl}}{51.48 \text{ g}} \times 100 = 68.87\% \text{ Cl}$$

e.   mass of P present =   30.97 g =   30.97 g

mass of H present = 3(1.008 g) = 3.024 g

molar mass of PH_3 =   33.99 g

$$\%P = \frac{30.97 \text{ g P}}{33.99 \text{ g}} \times 100 = 91.10\% \text{ P}$$

$$\%H = \frac{3.024 \text{ g H}}{33.99 \text{ g}} \times 100 = 8.896\% \text{ H}$$

f.   mass of H present = 3(1.008) g = 3.024 g

mass of P present =   30.97 g =   30.97 g

mass of O present = 3(16.00 g) = 48.00 g

molar mass of H_3PO_3 =   81.99 g

$$\%H = \frac{3.024 \text{ g H}}{81.99 \text{ g}} \times 100 = 3.688\% \text{ H}$$

$$\%P = \frac{30.97 \text{ g P}}{81.99 \text{ g}} \times 100 = 37.77\% \text{ P}$$

$$\%O = \frac{48.00 \text{ g O}}{81.99 \text{ g}} \times 100 = 58.54\% \text{ O}$$

48. a. molar mass of $CuBr_2$ = 223.4

$$\% \ Cu = \frac{63.55 \text{ g Cu}}{223.4 \text{ g}} \times 100 = 28.45\% \ Cu$$

   b. molar mass of CuBr = 143.5 g

$$\% \ Cu = \frac{63.55 \text{ g Cu}}{143.5 \text{ g}} \times 100 = 44.29\% \ Cu$$

   c. molar mass of $FeCl_2$ = 126.75 g

$$\% \ Fe = \frac{55.85 \text{ g Fe}}{126.75 \text{ g}} \times 100 = 44.06\% \ Fe$$

   d. molar mass of $FeCl_3$ = 162.2 g

$$\% \ Fe = \frac{55.85 \text{ g Fe}}{162.2 \text{ g}} \times 100 = 34.43\% \ Fe$$

   e. molar mass of CoI_2 = 312.7 g

$$\% \ Co = \frac{58.93 \text{ g Co}}{312.7 \text{ g}} \times 100 = 18.85\% \ Co$$

   f. molar mass of CoI_3 = 439.6 g

$$\% \ Co = \frac{58.93 \text{ g Co}}{439.6 \text{ g}} \times 100 = 13.41\% \ Co$$

   g. molar mass of SnO = 134.7 g

$$\% \ Sn = \frac{118.7 \text{ g Sn}}{134.7 \text{ g}} \times 100 = 88.12\% \ Sn$$

   h. molar mass of SnO_2 = 150.7 g

$$\% \ Sn = \frac{118.7 \text{ g Sn}}{150.7 \text{ g}} \times 100 = 78.77\% \ Sn$$

50. a. molar mass of $FeCl_3$ = 162.2 g

$$\% Fe = \frac{55.85 \text{ g Fe}}{162.2 \text{ g}} \times 100 = 34.43\% \ Fe$$

   b. molar mass of OF_2 = 54.00 g

$$\% O = \frac{16.00 \text{ g O}}{54.00 \text{ g}} \times 100 = 29.63\% \ O$$

   c. molar mass of C_6H_6 = 78.11 g

$$\% C = \frac{72.06 \text{ g C}}{78.11 \text{ g}} \times 100 = 92.25\% \ C$$

d.   molar mass of NH_4ClO_4 = 117.5 g

$$\%N = \frac{14.01 \text{ g N}}{117.5 \text{ g}} \times 100 = 11.92\% \text{ N}$$

e.   molar mass of Ag_2O = 231.8 g

$$\%Ag = \frac{215.8 \text{ g Ag}}{231.8 \text{ g}} \times 100 = 93.10\% \text{ Ag}$$

f.   molar mass of $CoCl_2$ = 129.83 g

$$\%Co = \frac{58.93 \text{ g Co}}{129.83 \text{ g}} \times 100 = 45.39\% \text{ Co}$$

g.   molar mass of N_2O_4 = 92.02 g

$$\%N = \frac{28.02 \text{ g N}}{92.02 \text{ g}} \times 100 = 30.45\% \text{ N}$$

h.   molar mass of $MnCl_2$ = 125.84 g

$$\%Mn = \frac{54.94 \text{ g Mn}}{125.8 \text{ g}} \times 100 = 43.66\% \text{ Mn}$$

52.   a.   molar mass of NH_4Cl = 53.49 g

molar mass of NH_4^+ ion = 18.04 g

$$\% NH_4^+ = \frac{18.04 \text{ g } NH_4^+}{53.49 \text{ g}} \times 100 = 33.73\% \ NH_4^+$$

b.   molar mass of $CuSO_4$ = 159.62 g

molar mass of Cu^{2+} = 63.55 g

$$\% Cu^{2+} = \frac{63.55 \text{ g } Cu^{2+}}{159.62 \text{ g}} \times 100 = 39.81\% \ Cu^{2+}$$

c.   molar mass of $AuCl_3$ = 303.4 g

molar mass of Au^{3+} ion = 197.0 g

$$\% Au^{3+} = \frac{197.0 \text{ g } Au^{3+}}{303.4 \text{ g}} \times 100 = 64.93\% \ Au^{3+}$$

d.   molar mass of $AgNO_3$ = 169.9 g

molar mass of Ag^+ ion = 107.9 g

$$\% Ag^+ = \frac{107.9 \text{ g } Ag^+}{169.9 \text{ g}} \times 100 = 63.51\% \ Ag^+$$

54.   The empirical formula represents the smallest whole number ratio of the elements present in a compound. The molecular formula indicates the actual number of atoms of each element found in a molecule of the substance.

56. a   yes (each of these has empirical formula CH)

    b.   no (the number of hydrogen atoms is wrong)

    c.   yes (both have empirical formula NO_2)

    d.   no (the number of hydrogen and oxygen atoms is wrong)

58.   Assume that we have 100.00 g of the compound: then 3.09 g will be hydrogen, 31.60 g will be phosphorus, and 65.31 g will be oxygen.

$$3.09 \text{ g H} \times \frac{1 \text{ mol H}}{1.008 \text{ g H}} = 3.07 \text{ mol H}$$

$$31.60 \text{ g P} \times \frac{1 \text{ mol P}}{30.97 \text{ g P}} = 1.020 \text{ mol P}$$

$$65.31 \text{ g P} \times \frac{1 \text{ mol O}}{16.00 \text{ g O}} = 4.082 \text{ mol O}$$

Dividing each number of moles by the smallest number of moles (1.020 mol P) gives

$$\frac{3.07 \text{ mol H}}{1.020} = 3.01 \text{ mol H}$$

$$\frac{1.020 \text{ mol P}}{1.020} = 1.000 \text{ mol P}$$

$$\frac{4.082 \text{ mol O}}{1.020} = 4.002 \text{ mol O}$$

The empirical formula is H_3PO_4.

60.   Consider having 100.0 g of the compound. Then the percentages of the elements present are numerically equal to their masses in grams.

$$58.84 \text{ g Ba} \times \frac{1 \text{ mol}}{137.3 \text{ g Ba}} = 0.4286 \text{ mol Ba}$$

$$13.74 \text{ g S} \times \frac{1 \text{ mol}}{32.07 \text{ g S}} = 0.4284 \text{ mol S}$$

$$27.43 \text{ g O} \times \frac{1 \text{ mol}}{16.00 \text{ g O+}} = 1.714 \text{ mol O}$$

Dividing each number of moles by the smallest number of moles (0.4284 mol S) gives

$$\frac{0.4286 \text{ mol Ba}}{0.4284} = 1.000 \text{ mol Ba}$$

$$\frac{0.4284 \text{ mol S}}{0.4284} = 1.000 \text{ mol S}$$

$$\frac{1.714 \text{ mol O}}{0.4284} = 4.001 \text{ mol O}$$

The empirical formula is $BaSO_4$.

62.   Consider 100.0 g of the compound.

$$29.16 \text{ g N} \times \frac{1 \text{ mol}}{14.01 \text{ g N}} = 2.081 \text{ mol N}$$

$$8.392 \text{ g H} \times \frac{1 \text{ mol}}{1.008 \text{ g H}} = 8.325 \text{ mol H}$$

$$12.50 \text{ g C} \times \frac{1 \text{ mol}}{12.01 \text{ g C}} = 1.041 \text{ mol C}$$

$$49.95 \text{ g O} \times \frac{1 \text{ mol}}{16.00 \text{ g O}} = 3.122 \text{ mol O}$$

Dividing each number of moles by the smallest (1.041 mol C) gives

$$\frac{2.081 \text{ mol N}}{1.041} = 1.999 \text{ mol N}$$

$$\frac{8.325 \text{ mol H}}{1.041} = 7.997 \text{ mol H}$$

$$\frac{1.041 \text{ mol C}}{1.041} = 1.000 \text{ mol C}$$

$$\frac{3.121 \text{ mol O}}{1.041} = 2.998 \text{ mol O}$$

The empirical formula is $N_2H_8CO_3$ [i.e, $(NH_4)_2CO_3$].

64.   Consider 100.0 g of the compound.

$$55.06 \text{ g Co} \times \frac{1 \text{ mol}}{58.93 \text{ g Co}} = 0.9343 \text{ mol Co}$$

If the sulfide of cobalt is 55.06% Co, then it is 44.94% S by mass.

$$44.94 \text{ g S} \times \frac{1 \text{ mol}}{32.07 \text{ g S}} = 1.401 \text{ mol S}$$

Dividing each number of moles by the smaller (0.9343 mol Co) gives

$$\frac{0.9343 \text{ mol Co}}{0.9343} = 1.000 \text{ mol Co}$$

$$\frac{1.401 \text{ mol S}}{0.9343} = 1.500 \text{ mol S}$$

Multiplying by two, to convert to whole numbers of moles, gives the empirical formula for the compound as Co_2S_3.

66. $10.00 \text{ g Cu} \times \dfrac{1 \text{ mol}}{63.55 \text{ g Cu}} = 0.1574 \text{ mol Cu}$

$2.52 \text{ g O} \times \dfrac{1 \text{ mol}}{16.00 \text{ g O}} = 0.158 \text{ mol O}$

The numbers of moles are almost equal: the empirical formula is just CuO.

68.   Consider 100.0 g of the compound.

$32.13 \text{ g Al} \times \dfrac{1 \text{ mol}}{26.98 \text{ g Al}} = 1.191 \text{ mol Al}$

$67.87 \text{ g F} \times \dfrac{1 \text{ mol}}{19.00 \text{ g F}} = 3.572 \text{ mol F}$

Dividing each number of moles by the smaller number (1.191 mol Al) gives

$$\frac{1.191 \text{ mol Al}}{1.191} = 1.000 \text{ mol Al}$$

$$\frac{3.572 \text{ mol F}}{1.191} = 2.999 \text{ mol F}$$

The empirical formula is AlF_3.

70.   Consider 100.0 g of the compound.

$59.78 \text{ g Li} \times \dfrac{1 \text{ mol Li}}{6.941 \text{ g Li}} = 8.613 \text{ mol Li}$

$40.22 \text{ g N} \times \dfrac{1 \text{ mol N}}{14.01 \text{ g N}} = 2.871 \text{ mol N}$

Dividing each number of moles by the smaller number of moles (2.871 mol N) gives

$\dfrac{8.613 \text{ mol Li}}{2.871} = 3.000 \text{ mol Li}$ $\dfrac{2.871 \text{ mol N}}{2.871} = 1.000 \text{ mol N}$

The empirical formula is Li_3N.

72.   Consider 100.0 g of the compound.

$$15.77 \text{ g Al} \times \frac{1 \text{ mol Al}}{26.98 \text{ g Al}} = 0.5845 \text{ mol Al}$$

$$28.11 \text{ g S} \times \frac{1 \text{ mol S}}{32.07 \text{ g S}} = 0.8765 \text{ mol S}$$

$$56.12 \text{ g O} \times \frac{1 \text{ mol O}}{16.00 \text{ g O}} = 3.508 \text{ mol O}$$

Dividing each number of moles by the smallest number of moles (0.5845 mol Al) gives

$$\frac{0.5845 \text{ mol Al}}{0.5845} = 1.000 \text{ mol Al}$$

$$\frac{0.8765 \text{ mol S}}{0.5845} = 1.500 \text{ mol S}$$

$$\frac{3.508 \text{ mol O}}{0.5845} = 6.002 \text{ mol O}$$

Multiplying these relative numbers of moles by 2 to give whole numbers gives the empirical formula as $Al_2S_3O_{12}$ [i.e., $Al_2(SO_4)_3$].

74.   Compound 1: Assume 100.0 g of the compound.

$$22.55 \text{ g P} \times \frac{1 \text{ mol P}}{30.97 \text{ g P}} = 0.7281 \text{ mol P}$$

$$77.45 \text{ g Cl} \times \frac{1 \text{ mol Cl}}{35.45 \text{ g Cl}} = 2.185 \text{ mol Cl}$$

Dividing each number of moles by the smaller (0.7281 mol P) indicates that the formula of Compound 1 is PCl_3.

Compound 2: Assume 100.0 g of the compound.

$$14.87 \text{ g P} \times \frac{1 \text{ mol P}}{30.97 \text{ g P}} = 0.4801 \text{ mol P}$$

$$85.13 \text{ g Cl} \times \frac{1 \text{ mol Cl}}{35.45 \text{ g Cl}} = 2.401 \text{ mol Cl}$$

Dividing each number of moles by the smaller (0.4801 mol P) indicates that the formula of Compound 2 is PCl_5.

76.   If only the empirical formula is known, the molar mass of the substance must be determined before the molecular formula can be calculated.

78.   empirical formula mass of CH = 13 g

$$n = \frac{\text{molar mass}}{\text{empirical formula mass}} = \frac{78 \text{ g}}{13 \text{ g}} = 6$$

The molecular formula is $(CH)_6$ or C_6H_6.

80.   empirical formula mass of CH_4O = 32.04 g

$$n = \frac{\text{molar mass}}{\text{empirical formula mass}} = \frac{192 \text{ g}}{32.04 \text{ g}} = 6$$

molecular formula is $(CH_4O)_6 = C_6H_{24}O_6$

82. $46.91 \text{ g Na} \times \dfrac{1 \text{ mol Na}}{22.99 \text{ g Na}} = 2.040 \text{ mol Na}$

$24.51 \text{ g C} \times \dfrac{1 \text{ mol C}}{12.01 \text{ g C}} = 2.041 \text{ mol C}$

$28.59 \text{ g N} \times \dfrac{1 \text{ mol N}}{14.01 \text{ g N}} = 2.041$

The empirical formula is NaCN. The empirical molar mass would be (22.99 g + 12.01 g + 14.01 g) = 49.01 g. Since the experimental molar mass is approximately 50 g, then the molecular formula must also be NaCN.

84.

5.00 g Al	0.185 mol	1.12×10^{23} atoms
0.140 g Fe	0.00250 mol	1.51×10^{21} atoms
2.7×10^2 g Cu	4.3 mol	2.6×10^{24} atoms
0.00250 g Mg	1.03×10^{-4} mol	6.19×10^{19} atoms
0.062 g Na	2.7×10^{-3} mol	1.6×10^{21} atoms
3.95×10^{-18} g U	1.66×10^{-20} mol	1.00×10^4 atoms

86.   mass of 2 mol X =   2(41.2 g) =   82.4 g

mass of 1 mol Y =   57.7 g =   57.7 g

mass of 3 mol Z =   3(63.9 g) =   191.7 g

molar mass of X_2YZ_3 =   331.8 g

$$\% \text{ X} = \frac{82.4 \text{ g}}{331.8 \text{ g}} \times 100 = 24.8\% \text{ X}$$

$$\% \text{ Y} = \frac{57.7 \text{ g}}{331.8 \text{ g}} \times 100 = 17.4\% \text{ Y}$$

$$\% \ Z = \frac{191.7 \ g}{331.8 \ g} \times 100 = 57.8\% \quad Z$$

If the molecular formula were actually $X_4Y_2Z_6$, the percentage composition would be the same, and the *relative* mass of each element present would not change. The molecular formula is always a whole number multiple of the empirical formula.

88.   For the first compound (*restricted* amount of oxygen)

$$2.118 \ g \ Cu \times \frac{1 \ mol \ Cu}{63.54 \ g \ Cu} = 0.03333 \ mol \ Cu$$

$$0.2666 \ g \ O \times \frac{1 \ mol \ O}{16.00 \ g \ O} = 0.01666 \ mol \ O$$

Since the number of moles of Cu (0.03333 mol) is twice the number of moles of O (0.01666 mol), the empirical formula is Cu_2O.

For the second compound (stream of pure oxygen)

$$2.118 \ g \ Cu \times \frac{1 \ mol \ Cu}{63.54 \ g \ Cu} = 0.03333 \ mol \ Cu$$

$$0.5332 \ g \ O \times \frac{1 \ mol \ O}{16.00 \ g \ O} = 0.03333 \ mol \ O$$

Since the numbers of moles are the same, the empirical formula is CuO.

90.   a.   molar mass H_2O = 18.02 g

$$4.21 \ g \times \frac{1 \ mol}{18.02 \ g} \times \frac{6.022 \times 10^{23} \ molecules}{1 \ mol} = 1.41 \times 10^{23} \ molecules$$

The sample contains 1.41×10^{23} oxygen atoms and $2(1.41 \times 10^{23})$ = 2.82×10^{23} hydrogen atoms.

b.   molar mass CO_2 = 44.01 g

$$6.81 \ g \times \frac{1 \ mol}{44.01 \ g} \times \frac{6.022 \times 10^{23} \ molecules}{1 \ mol} = 9.32 \times 10^{22} \ molecules$$

The sample contains 9.32×10^{22} carbon atoms and $2(9.32 \times 10^{22})$ = 1.86×10^{23} oxygen atoms.

c.   molar mass C_6H_6 = 78.11 g

$$0.000221 \ g \times \frac{1 \ mol}{78.11 \ g} \times \frac{6.022 \times 10^{23} \ molecules}{1 \ mol} = 1.70 \times 10^{18} \ molec.$$

The sample contains $6(1.70 \times 10^{18})$ = 1.02×10^{19} atoms of each element.

d. $2.26 \text{ mol} \times \dfrac{6.022 \times 10^{23} \text{ molecules}}{1 \text{ mol}} = 1.36 \times 10^{24} \text{ molecules}$

atoms C = $12(1.36 \times 10^{24}) = 1.63 \times 10^{25}$ atoms

atoms H = $22(1.36 \times 10^{24}) = 2.99 \times 10^{25}$ atoms

atoms O = $11(1.36 \times 10^{24}) = 1.50 \times 10^{25}$ atoms

92.   a.   molar mass of C_3O_2 = 3(12.01 g) + 2(16.00 g) = 68.03 g

$\% \text{ C} = \dfrac{36.03 \text{ g}}{68.03 \text{ g}} = 52.96\% \text{ C}$

$7.819 \text{ g } C_3O_2 \times \dfrac{52.96 \text{ g C}}{100.0 \text{ g } C_3O_2} = 4.141 \text{ g C}$

$4.141 \text{ g C} \times \dfrac{6.022 \times 10^{23} \text{ C atoms}}{12.01 \text{ g C}} = 2.076 \times 10^{23} \text{ C atoms}$

b.   molar mass of CO = 12.01 g + 16.00 g = 28.01 g

$\% \text{ C} = \dfrac{12.01 \text{ g}}{28.01 \text{ g}} \times 100 = 42.88\% \text{ C}$

$1.53 \times 10^{21} \text{ molecules CO} \times \dfrac{1 \text{ C atom}}{1 \text{ molecule CO}} = 1.53 \times 10^{21} \text{ C atoms}$

$1.53 \times 10^{21} \text{ C atoms} \times \dfrac{12.01 \text{ g C}}{6.022 \times 10^{23} \text{ C atoms}} = 0.0305 \text{ g C}$

c.   molar mass of C_6H_6O = 6(12.01 g) + 6(1.008 g) + 16.00 g = 94.11 g

$\% \text{ C} = \dfrac{72.06 \text{ g}}{94.11 \text{ g}} \times 100 = 76.57\% \text{ C}$

$0.200 \text{ mol } C_6H_6O \times \dfrac{6 \text{ mol C}}{1 \text{ mol } C_6H_6O} = 1.20 \text{ mol C}$

$1.20 \text{ mol C} \times \dfrac{12.01 \text{ g C}}{1 \text{ mol C}} = 14.4 \text{ g C}$

$14.4 \text{ g C} \times \dfrac{6.022 \times 10^{23} \text{ C atoms}}{12.01 \text{ g C}} = 7.22 \times 10^{23} \text{ C atoms}$

94. $2.24 \text{ g Co} \times \dfrac{55.85 \text{ g Fe}}{58.93 \text{ g Co}} = 2.12 \text{ g Fe}$

96. $5.00 \text{ g Te} \times \dfrac{200.6 \text{ g Hg}}{127.6 \text{ g Te}} = 7.86 \text{ g Hg}$

98.   153.8 g CCl_4 = 6.022 × 10^{23} molecules CCl_4

$$1 \text{ molecule} \times \frac{153.8 \text{ g}}{6.022 \times 10^{23} \text{ molecules}} = 2.554 \times 10^{-22} \text{ g}$$

100.   a.   molar mass of $C_2H_5O_2N$ = 75.07 g

$$\text{mass fraction N} = \frac{14.01 \text{ g}}{75.07 \text{ g}}$$

$$5.000 \text{ g} \times \frac{14.01 \text{ g}}{75.07 \text{ g}} = 0.9331 \text{ g N}$$

b.   molar mass of Mg_3N_2 = 3(24.31 g) + 2(14.01 g) = 100.95 g

$$\text{mass fraction N} = \frac{28.02 \text{ g}}{100.95 \text{ g}}$$

$$5.000 \text{ g} \times \frac{28.02 \text{ g}}{100.95 \text{ g}} = 1.388 \text{ g N}$$

c.   molar mass of $Ca(NO_3)_2$ =

40.08 g + 2(14.01 g) + 6(16.00 g) = 164.10 g

$$\text{mass fraction N} = \frac{28.02 \text{ g}}{164.10 \text{ g}}$$

$$5.000 \text{ g} \times \frac{28.02 \text{ g}}{164.10 \text{ g}} = 0.8537 \text{ g N}$$

d.   molar mass of N_2O_4 = 2(14.01 g) + 4(16.00 g) = 92.02 g

$$\text{mass fraction N} = \frac{28.02 \text{ g}}{92.02 \text{ g}}$$

$$5.000 \text{ g} \times \frac{28.02 \text{ g}}{92.02 \text{ g}} = 1.522 \text{ g N}$$

102.   Consider 100.0 g of the compound.

$$16.39 \text{ g Mg} \times \frac{1 \text{ mol Mg}}{24.31 \text{ g Mg}} = 0.6742 \text{ mol Mg}$$

$$18.89 \text{ g N} \times \frac{1 \text{ mol N}}{14.01 \text{ g N}} = 1.348 \text{ mol N}$$

$$64.72 \text{ g O} \times \frac{1 \text{ mol O}}{16.00 \text{ g O}} = 4.045 \text{ mol O}$$

Dividing each number of moles by the smallest number of moles

$$\frac{0.6742 \text{ mol Mg}}{0.6742} = 1.000 \text{ mol Mg}$$

$$\frac{1.348 \text{ mol N}}{0.6742} = 1.999 \text{ mol N}$$

$$\frac{4.045 \text{ mol O}}{0.6742} = 5.999 \text{ mol O}$$

The empirical formula is MgN_2O_6 [i.e., $Mg(NO_3)_2$].

104.   We use the *average* mass because this average is a *weighted average* and takes into account both the masses and the relative abundances of the various isotopes.

106. 1.98×10^{13} amu $\times \dfrac{1 \text{ Na atom}}{22.99 \text{ amu}} = 8.61 \times 10^{11}$ Na atoms

3.01×10^{23} Na atoms $\times \dfrac{22.99 \text{ amu}}{1 \text{ Na atom}} = 6.92 \times 10^{24}$ amu

108.   a.   5.0 mol K $\times \dfrac{39.10 \text{ g}}{1 \text{ mol}} = 195 \text{ g} = 2.0 \times 10^2$ g K

   b.   0.000305 mol Hg $\times \dfrac{200.6 \text{ g}}{1 \text{ mol}} = 0.0612$ g Hg

   c. 2.31×10^{-5} mol Mn $\times \dfrac{54.94 \text{ g}}{1 \text{ mol}} = 1.27 \times 10^{-3}$ g Mn

   d.   10.5 mol P $\times \dfrac{30.97 \text{ g}}{1 \text{ mol}} = 325$ g P

   e. 4.9×10^4 mol Fe $\times \dfrac{55.85 \text{ g}}{1 \text{ mol}} = 2.7 \times 10^6$ g Fe

   f.   125 mol Li $\times \dfrac{6.941 \text{ g}}{1 \text{ mol}} = 868$ g Li

   g.   0.01205 mol F $\times \dfrac{19.00 \text{ g}}{1 \text{ mol}} = 0.2290$ g F

110.   a.   mass of 1 mol Fe =   1(55.85 g) =   55.85 g

   mass of 1 mol S =   1(32.07 g) =   32.07 g

   mass of 4 mol O =   4(16.00 g) =   64.00 g
   _____

   molar mass of $FeSO_4$ =   151.92 g

   b.   mass of 1 mol Hg = 1(200.6 g) =   200.6 g

   mass of 2 mol I =   2(126.9 g) =   253.8 g
   _____

   molar mass of HgI_2 =   454.4 g

c. mass of 1 mol Sn = 1(118.7 g) = 118.7 g

mass of 2 mol O =   2(16.00 g) = 32.00 g

molar mass of SnO_2 =   150.7 g

d. mass of 1 mol Co = 1(58.93 g) = 58.93 g

mass of 2 mol Cl = 2(35.45 g) = 70.90 g

molar mass of $CoCl_2$ =   129.83 g

e. mass of 1 mol Cu = 1(63.55 g) = 63.55 g

mass of 2 mol N =   2(14.01 g) = 28.02 g

mass of 6 mol O =   6(16.00 g) = 96.00 g

molar mass of $Cu(NO_3)_2$ =   187.57 g

112. a. molar mass of $(NH_4)_2S$ = 68.15 g

$$21.2 \text{ g} \times \frac{1 \text{ mol}}{68.15 \text{ g}} = 0.311 \text{ mol } (NH_4)_2S$$

b. molar mass of $Ca(NO_3)_2$ = 164.1 g

$$44.3 \text{ g} \times \frac{1 \text{ mol}}{164.1 \text{ g}} = 0.270 \text{ mol } Ca(NO_3)_2$$

c. molar mass of Cl_2O = 86.9 g

$$4.35 \text{ g} \times \frac{1 \text{ mol}}{86.9 \text{ g}} = 0.0501 \text{ mol } Cl_2O$$

d. 1.0 lb = 454 g

molar mass of $FeCl_3$ = 162.2

$$454 \text{ g} \times \frac{1 \text{ mol}}{162.2 \text{ g}} = 2.8 \text{ mol } FeCl_3$$

e. 1.0 kg = 1.0×10^3 g

molar mass of $FeCl_3$ = 162.2 g

$$1.0 \times 10^3 \text{ g} \times \frac{1 \text{ mol}}{162.2 \text{ g}} = 6.2 \text{ mol } FeCl_3$$

114. a. molar mass of $CuSO_4$ = 159.62 g

$$2.6 \times 10^{-2} \text{ mol} \times \frac{159.62 \text{ g}}{1 \text{ mol}} = 4.2 \text{ g } CuSO_4$$

b.   molar mass of C_2F_4 = 100.0 g

$$3.05 \times 10^3 \text{ mol} \times \frac{100.0 \text{ g}}{1 \text{ mol}} = 3.05 \times 10^5 \text{ g } C_2F_4$$

c.   7.83 mmol = 0.00783 mol

molar mass of C_5H_8 = 68.11 g

$$0.00783 \text{ mol} \times \frac{68.11 \text{ g}}{1 \text{ mol}} = 0.533 \text{ g } C_5H_8$$

d.   molar mass of $BiCl_3$ = 315.3 g

$$6.30 \text{ mol} \times \frac{315.3 \text{ g}}{1 \text{ mol}} = 1.99 \times 10^3 \text{ g } BiCl_3$$

e.   molar mass of $C_{12}H_{22}O_{11}$ = 342.3 g

$$12.2 \text{ mol} \times \frac{342.3 \text{ g}}{1 \text{ mol}} = 4.18 \times 10^3 \text{ g } C_{12}H_{22}O_{11}$$

116.   a.   molar mass of $C_6H_{12}O_6$ = 180.2 g

$$3.45 \text{ g} \times \frac{6.022 \times 10^{23} \text{ molecules}}{180.2 \text{ g}} = 1.15 \times 10^{22} \text{ molecules } C_6H_{12}O_6$$

b. $$3.45 \text{ mol} \times \frac{6.022 \times 10^{23} \text{ molecules}}{1 \text{ mol}} = 2.08 \times 10^{24} \text{ molecules } C_6H_{12}O_6$$

c.   molar mass of ICl_5 = 304.2 g

$$25.0 \text{ g} \times \frac{6.022 \times 10^{23} \text{ molecules}}{304.2 \text{ g}} = 4.95 \times 10^{22} \text{ molecules } ICl_5$$

d.   molar mass of B_2H_6 = 27.67 g

$$1.00 \text{ g} \times \frac{6.022 \times 10^{23} \text{ molecules}}{27.67 \text{ g}} = 2.18 \times 10^{22} \text{ molecules } B_2H_6$$

e.   1.05 mmol = 0.00105 mol

$$0.00105 \text{ mol} \times \frac{6.022 \times 10^{23} \text{ form. units}}{1 \text{ mol}} = 6.32 \times 10^{20} \text{ form. units}$$

118.   a.   mass of Ca present =   3(40.08 g) =   120.24 g

mass of P present =   2(30.97 g) =   61.94 g

mass of O present =   8(16.00 g) =   128.00 g

molar mass of $Ca_3(PO_4)_2$ =   310.18 g

$$\% \text{ Ca} = \frac{120.24 \text{ g Ca}}{310.18 \text{ g}} \times 100 = 38.76\% \text{ Ca}$$

$$\% \text{ P} = \frac{61.94 \text{ g P}}{310.18 \text{ g}} \times 100 = 19.97\% \text{ P}$$

$$\% \text{ O} = \frac{128.00 \text{ g O}}{310.18 \text{ g}} \times 100 = 41.27\% \text{ O}$$

b.   mass of Cd present = 112.4 g =   112.4 g
     mass of S present =   32.07 g =   32.07 g
     mass of O present =   4(16.00 g) =   64.00 g
     ───
     molar mass of $CdSO_4$ =   208.5 g

$$\% \text{ Cd} = \frac{112.4 \text{ g Cd}}{208.5 \text{ g}} \times 100 = 53.91\% \text{ Cd}$$

$$\% \text{ S} = \frac{32.07 \text{ g S}}{208.5 \text{ g}} \times 100 = 15.38\% \text{ S}$$

$$\% \text{ O} = \frac{64.00 \text{ g O}}{208.5 \text{ g}} \times 100 = 30.70\% \text{ O}$$

c.   mass of Fe present =   2(55.85 g) = 111.7 g
     mass of S present =   3(32.07 g) =   96.21 g
     mass of O present =   12(16.00 g) = 192.0 g
     ───
     molar mass of $Fe_2(SO_4)_3$ =   399.9 g

$$\% \text{ Fe} = \frac{111.7 \text{ g Fe}}{399.9 \text{ g}} \times 100 = 27.93\% \text{ Fe}$$

$$\% \text{ S} = \frac{96.21 \text{ g S}}{399.9 \text{ g}} \times 100 = 24.06\% \text{ S}$$

$$\% \text{ O} = \frac{192.0 \text{ g O}}{399.9 \text{ g}} \times 100 = 48.01\% \text{ O}$$

d.   mass of Mn present = 54.94 g =   54.94 g
     mass of Cl present = 2(35.45 g) =   70.90 g
     ───
     molar mass of $MnCl_2$ =   125.84 g

$$\% \text{ Mn} = \frac{54.94 \text{ g Mn}}{125.84 \text{ g}} \times 100 = 43.66\% \text{ Mn}$$

$$\% \text{ Cl} = \frac{70.90 \text{ g Cl}}{125.84 \text{ g}} \times 100 = 56.34\% \text{ Cl}$$

e.   mass of N present =   2(14.01 g) =   28.02 g

   mass of H present =   8(1.008 g) =   8.064 g

   mass of C present =   12.01 g =   12.01 g

   mass of O present =   3(16.00 g) =   48.00 g
   _____

   molar mass of $(NH_4)_2CO_3$ =   96.09 g

   % N = $\dfrac{28.02 \text{ g N}}{96.09 \text{ g}}$ × 100 = 29.16% N

   % H = $\dfrac{8.064 \text{ g H}}{96.09 \text{ g}}$ × 100 = 8.392% H

   % C = $\dfrac{12.01 \text{ g C}}{96.09 \text{ g}}$ × 100 = 12.50% C

   % O = $\dfrac{48.00 \text{ g O}}{96.09 \text{ g}}$ × 100 = 49.95% O

f.   mass of Na present =   22.99 g =   22.99 g

   mass of H present =   1.008 g =   1.008 g

   mass of C present =   12.01 g =   12.01 g

   mass of O present =   3(16.00 g) =   48.00 g
   _____

   molar mass of $NaHCO_3$ =   84.01 g

   % Na = $\dfrac{22.99 \text{ g Na}}{84.01 \text{ g}}$ × 100 = 27.37% Na

   % H = $\dfrac{1.008 \text{ g H}}{84.01 \text{ g}}$ × 100 = 1.200% H

   % C = $\dfrac{12.01 \text{ g C}}{84.01 \text{ g}}$ × 100 = 14.30% C

   % O = $\dfrac{48.00 \text{ g O}}{84.01 \text{ g}}$ × 100 = 57.14% O

g.   mass of C present =   12.01 g =   12.01 g

   mass of O present =   2(16.00 g) =   32.00 g
   _____

   molar mass of CO_2 =   44.01 g

   % C = $\dfrac{12.01 \text{ g C}}{44.01 \text{ g}}$ × 100 = 27.29% C

   % O = $\dfrac{32.00 \text{ g O}}{44.01 \text{ g}}$ × 100 = 72.71% O

h.   mass of Ag present = 107.9 g =   107.9 g

mass of N present =   14.01 g =   14.01 g

mass of O present =   3(16.00 g) =   48.00 g
_____

molar mass of $AgNO_3$ =   169.9 g

$$\% \text{ Ag} = \frac{107.9 \text{ g Ag}}{169.9 \text{ g}} \times 100 = 63.51\% \text{ Ag}$$

$$\% \text{ N} = \frac{14.01 \text{ g N}}{169.9 \text{ g}} \times 100 = 8.246\% \text{ N}$$

$$\% \text{ O} = \frac{48.00 \text{ g O}}{169.9 \text{ g}} \times 100 = 28.25\% \text{ O}$$

120. a. $$\% \text{ Fe} = \frac{55.85 \text{ g Fe}}{151.92 \text{ g}} \times 100 = 36.76\% \text{ Fe}$$

b. $$\% \text{ Ag} = \frac{215.8 \text{ g Ag}}{231.8 \text{ g}} \times 100 = 93.10\% \text{ Ag}$$

c. $$\% \text{ Sr} = \frac{87.62 \text{ g Sr}}{158.5 \text{ g}} \times 100 = 55.28\% \text{ Sr}$$

d. $$\% \text{ C} = \frac{48.04 \text{ g C}}{86.09 \text{ g}} \times 100 = 55.80\% \text{ C}$$

e. $$\% \text{ C} = \frac{12.01 \text{ g C}}{32.04 \text{ g}} \times 100 = 37.48\% \text{ C}$$

f. $$\% \text{ Al} = \frac{53.96 \text{ g Al}}{101.96 \text{ g}} \times 100 = 52.92\% \text{ Al}$$

g. $$\% \text{ K} = \frac{39.10 \text{ g K}}{106.55 \text{ g}} \times 100 = 36.70\% \text{ K}$$

h. $$\% \text{ K} = \frac{39.10 \text{ g K}}{74.55 \text{ g}} \times 100 = 52.45\% \text{ K}$$

122. $$0.2990 \text{ g C} \times \frac{1 \text{ mol C}}{12.01 \text{ g C}} = 0.02490 \text{ mol C}$$

$$0.05849 \text{ g H} \times \frac{1 \text{ mol H}}{1.008 \text{ g H}} = 0.05803 \text{ mol H}$$

$$0.2318 \text{ g N} \times \frac{1 \text{ mol N}}{14.01 \text{ g N}} = 0.01655 \text{ mol N}$$

$$0.1328 \text{ g O} \times \frac{1 \text{ mol O}}{16.00 \text{ g O}} = 0.008300 \text{ mol O}$$

Dividing each number of moles by the smallest number of moles (0.008300 mol O) gives

$$\frac{0.02490 \text{ mol C}}{0.008300} = 3.000 \text{ mol C}$$

$$\frac{0.05803 \text{ mol H}}{0.008300} = 6.992 \text{ mol H}$$

$$\frac{0.01655 \text{ mol N}}{0.008300} = 1.994 \text{ mol N}$$

$$\frac{0.008300 \text{ mol O}}{0.008300} = 1.000 \text{ mol O}$$

The empirical formula is $C_3H_7N_2O$.

124. Mass of oxygen in compound = 4.33 g – 4.01 g = 0.32 g O

$$4.01 \text{ g Hg} \times \frac{1 \text{ mol Hg}}{200.6 \text{ g Hg}} = 0.0200 \text{ mol Hg}$$

$$0.32 \text{ g O} \times \frac{1 \text{ mol O}}{16.00 \text{ g O}} = 0.020 \text{ mol O}$$

Since the numbers of moles are equal, the empirical formula is HgO.

126. Assume we have 100.0 g of the compound.

$$65.95 \text{ g Ba} \times \frac{1 \text{ mol Ba}}{137.3 \text{ g Ba}} = 0.4803 \text{ mol Ba}$$

$$34.05 \text{ g Cl} \times \frac{1 \text{ mol Cl}}{35.45 \text{ g Cl}} = 0.9605 \text{ mol Cl}$$

Dividing each of these number of moles by the smaller number gives

$$\frac{0.4803 \text{ mol Ba}}{0.4803} = 1.000 \text{ mol Ba}$$

$$\frac{0.9605 \text{ mol Cl}}{0.4803} = 2.000 \text{ mol Cl}$$

The empirical formula is then $BaCl_2$.

Chapter 9   Chemical Quantities

2. The coefficients of the balanced chemical equation for a reaction
    indicate the *relative numbers of moles* of each reactant that combine
    during the process, as well as the number of moles of each product
    formed.

4. Balanced chemical equations tell us in what proportions *on a mole basis*
    substances combine; since the molar masses of $C(s)$ and $O_2(g)$ are
    different, 1 g of O_2 could not represent the same number of moles as 1 g
    of C.

6. a. $3MnO_2(s) + 4Al(s) \rightarrow 3Mn(s) + 2Al_2O_3(s)$

          Three formula units of manganese(IV) oxide react with four
          aluminum atoms, producing three manganese atoms and two formula
          units of aluminum oxide. Three moles of solid manganese(IV) oxide
          react with four moles of solid aluminum, to produce three moles of
          solid manganese and two moles of solid aluminum oxide.

    b. $B_2O_3(s) + 3CaF_2(s) \rightarrow 2BF_3(g) + 3CaO(s)$

          One molecule of diboron trioxide reacts with three formula units
          of calcium fluoride, producing two molecules of boron trifluoride
          and three formula units of calcium oxide. One mole of solid
          diboron trioxide reacts with three moles of solid calcium
          fluoride, to give two moles of gaseous boron trifluoride and three
          moles of solid calcium oxide.

    c. $3NO_2(g) + H_2O(l) \rightarrow 2HNO_3(aq) + NO(g)$

          Three molecules of nitrogen dioxide [nitrogen(IV) oxide] react
          with one molecule of water, to produce two molecules of nitric
          acid and one molecule of nitrogen monoxide [nitrogen(II) oxide].
          Three moles of gaseous nitrogen dioxide react with one mole of
          liquid water, to produce two moles of aqueous nitric acid and one
          mole of nitrogen monoxide gas.

    d. $C_6H_6(g) + 3H_2(g) \rightarrow C_6H_{12}(g)$

          One molecule of C_6H_6 (which is named benzene) reacts with three
          molecules of hydrogen, producing just one molecule of C_6H_{12} (which
          is named cyclohexane). One mole of gaseous benzene reacts with
          three moles of hydrogen gas, giving one mole of gaseous
          cyclohexane.

8. False. Reactions take place on a *mole* basis: one *mole* of nitrogen gas
    would react with three *moles* of iodine, giving two *moles* of nitrogen
    triiodide.

10. $CaH_2(g) + 2H_2O(l) \rightarrow Ca(OH)_2(aq) + 2H_2(g)$

$$\frac{1 \text{ mol } Ca(OH)_2}{2 \text{ mol } H_2O} \quad \text{and} \quad \frac{2 \text{ mol } H_2}{2 \text{ mol } H_2O}$$

12.   a. $C_{12}H_{22}O_{11}(s) \rightarrow 12C(s) + 11H_2O(g)$

$$0.250 \text{ mol } C_{12}H_{22}O_{11} \times \frac{12 \text{ mol } C}{1 \text{ mol } C_{12}H_{22}O_{11}} = 3.00 \text{ mol } C$$

$$0.250 \text{ mol } C_{12}H_{22}O_{11} \times \frac{11 \text{ mol } H_2O}{1 \text{ mol } C_{12}H_{22}O_{11}} = 2.75 \text{ mol } H_2O$$

b. $S(s) + 2H_2SO_4(aq) \rightarrow 3SO_2(g) + 2H_2O(l)$

$$0.250 \text{ mol } S \times \frac{3 \text{ mol } SO_2}{1 \text{ mol } S} = 0.750 \text{ mol } SO_2$$

$$0.250 \text{ mol } S \times \frac{2 \text{ mol } H_2O}{1 \text{ mol } S} = 0.500 \text{ mol } H_2O$$

c. $2Bi_2O_3(s) \rightarrow 4Bi(s) + 3O_2(g)$

$$0.250 \text{ mol } Bi_2O_3 \times \frac{4 \text{ mol } Bi}{2 \text{ mol } Bi_2O_3} = 0.500 \text{ mol } Bi$$

$$0.250 \text{ mol } Bi_2O_3 \times \frac{3 \text{ mol } O_2}{2 \text{ mol } Bi_2O_3} = 0.375 \text{ mol } H_2O$$

d. $2H_2S(g) + 3O_2(g) \rightarrow 2SO_2(g) + 2H_2O(l)$

$$0.250 \text{ mol } H_2S \times \frac{2 \text{ mol } SO_2}{2 \text{ mol } H_2S} = 0.250 \text{ mol } SO_2$$

$$0.250 \text{ mol } H_2S \times \frac{2 \text{ mol } H_2O}{2 \text{ mol } H_2S} = 0.250 \text{ mol } H_2O$$

14.   a. $NH_3(g) + HCl(g) \rightarrow NH_4Cl(s)$

molar mass of NH_4Cl, 53.49 g

$$0.50 \text{ mol } NH_3 \times \frac{1 \text{ mol } NH_4Cl}{1 \text{ mol } NH_3} = 0.50 \text{ mol } NH_4Cl$$

$$0.50 \text{ mol } NH_4Cl \times \frac{53.49 \text{ g } NH_4Cl}{1 \text{ mol } NH_4Cl} = 27 \text{ g } NH_4Cl$$

b. $CH_4(g) + 4S(g) \rightarrow CS_2(l) + 2H_2S(g)$

molar masses: CS_2, 76.15 g; H_2S, 34.09 g

$$0.50 \text{ mol } S \times \frac{1 \text{ mol } CS_2}{4 \text{ mol } S} = 0.125 \text{ mol } CS_2 \ (= 0.13 \text{ mol } CS_2)$$

$$0.125 \text{ mol } CS_2 \times \frac{76.15 \text{ g } CS_2}{1 \text{ mol } CS_2} = 9.5 \text{ g } CS_2$$

$$0.50 \text{ mol } S \times \frac{2 \text{ mol } H_2S}{4 \text{ mol } S} = 0.25 \text{ mol } H_2S$$

$$0.25 \text{ mol } H_2S \times \frac{34.09 \text{ g } H_2S}{1 \text{ mol } H_2S} = 8.5 \text{ g } H_2S$$

c. $PCl_3(l) + 3H_2O(l) \rightarrow H_3PO_3(aq) + 3HCl(aq)$

molar masses: H_3PO_3, 81.99 g; HCl, 36.46 g

$$0.50 \text{ mol } PCl_3 \times \frac{1 \text{ mol } H_3PO_3}{1 \text{ mol } PCl_3} = 0.50 \text{ mol } H_3PO_3$$

$$0.50 \text{ mol } H_3PO_3 \times \frac{81.99 \text{ g } H_3PO_3}{1 \text{ mol } H_3PO_3} = 41 \text{ g } H_3PO_3$$

$$0.50 \text{ mol } PCl_3 \times \frac{3 \text{ mol } HCl}{1 \text{ mol } PCl_3} = 1.5 \text{ mol } HCl$$

$$1.5 \text{ mol } HCl \times \frac{36.46 \text{ g } HCl}{1 \text{ mol } HCl} = 54.7 = 55 \text{ g } HCl$$

d. $NaOH(s) + CO_2(g) \rightarrow NaHCO_3(s)$

molar mass of $NaHCO_3$ = 84.01 g

$$0.50 \text{ mol } NaOH \times \frac{1 \text{ mol } NaHCO_3}{1 \text{ mol } NaOH} = 0.50 \text{ mol } NaHCO_3$$

$$0.50 \text{ mol } NaHCO_3 \times \frac{84.01 \text{ g } NaHCO_3}{1 \text{ mol } NaHCO_3} = 42 \text{ g } NaHCO_3$$

16.   Before doing the calculations, the equations must be *balanced*.

a. $4KO_2(s) + 2H_2O(l) \rightarrow 3O_2(g) + 4KOH(s)$

$$0.625 \text{ mol } KOH \times \frac{3 \text{ mol } O_2}{4 \text{ mol } KOH} = 0.469 \text{ mol } O_2$$

b. $SeO_2(g) + 2H_2Se(g) \rightarrow 3Se(s) + 2H_2O(g)$

$$0.625 \text{ mol } H_2O \times \frac{3 \text{ mol } Se}{2 \text{ mol } H_2O} = 0.938 \text{ mol } Se$$

c. $2CH_3CH_2OH(l) + O_2(g) \rightarrow 2CH_3CHO(aq) + 2H_2O(l)$

$$0.625 \text{ mol } H_2O \times \frac{2 \text{ mol } CH_3CHO}{2 \text{ mol } H_2O} = 0.625 \text{ mol } CH_3CHO$$

d. $Fe_2O_3(s) + 2Al(s) \rightarrow 2Fe(l) + Al_2O_3(s)$

$$0.625 \text{ mol } Al_2O_3 \times \frac{2 \text{ mol } Fe}{1 \text{ mol } Al_2O_3} = 1.25 \text{ mol } Fe$$

18.   Stoichiometry is the process of using a chemical equation to calculate the relative masses of reactants and products involved in a reaction.

20.   a.   molar mass of Ir = 192.2 g

$$1.47 \times 10^{-3} \text{ g Ir} \times \frac{1 \text{ mol}}{192.2 \text{ g}} = 7.65 \times 10^{-6} \text{ mol Ir}$$

b.   molar mass of PbS = 239.3 g

$$8.95 \text{ g PbS} \times \frac{1 \text{ mol}}{239.3 \text{ g}} = 0.0374 \text{ mol PbS}$$

c.   molar mass of CuO = 79.55 g;   293 mg = 0.293 g

$$0.293 \text{ g CuO} \times \frac{1 \text{ mol}}{79.55} = 0.00368 \text{ mol CuO}$$

d.   molar mass of $FeCl_3$ = 162.2 g

$$91.4 \text{ g FeCl}_3 \times \frac{1 \text{ mol}}{162.2 \text{ g}} = 0.564 \text{ mol FeCl}_3$$

e.   molar mass of N_2 = 28.02 g

$$2.67 \text{ g N}_2 \times \frac{1 \text{ mol}}{28.02 \text{ g}} = 0.0953 \text{ mol N}_2$$

f.   molar mass of CS_2 = 76.15 g

$$89.2 \text{ g CS}_2 \times \frac{1 \text{ mol}}{76.15 \text{ g}} = 1.17 \text{ mol CS}_2$$

g.   molar mass of Fe = 55.85 g;   1.43 kg = 1430 g

$$1430 \text{ g Fe} \times \frac{1 \text{ mol}}{55.85 \text{ g}} = 25.6 \text{ mol Fe}$$

22.   a.   molar mass of CuI = 190.5 g

$$0.624 \text{ mol CuI} \times \frac{190.5 \text{ g}}{1 \text{ mol}} = 119 \text{ g CuI}$$

b.   molar mass of Br_2 = 159.8 g

$$4.24 \text{ mol Br}_2 \times \frac{159.8 \text{ g}}{1 \text{ mol}} = 678 \text{ g Br}_2$$

c.   molar mass of XeF_4 = 207.4 g

$$0.000211 \text{ mol XeF}_4 \times \frac{207.4 \text{ g}}{1 \text{ mol}} = 0.437 \text{ g XeF}_4$$

d.   molar mass of C_2H_4 = 28.05 g

$$9.11 \text{ mol C}_2\text{H}_4 \times \frac{28.05 \text{ g}}{1 \text{ mol}} = 256 \text{ g C}_2\text{H}_4$$

e.   molar mass of NH_3 = 17.03 g;   1.21 millimol = 0.00121 mol

$$0.00121 \text{ mol } NH_3 \times \frac{17.03 \text{ g}}{1 \text{ mol}} = 0.0206 \text{ g } NH_3$$

f.   molar mass of NaOH = 40.00 g

$$4.25 \text{ mol NaOH} \times \frac{40.00 \text{ g}}{1 \text{ mol}} = 170. \text{ g NaOH}$$

g.   molar mass of KI = 166.0 g

$$1.27 \times 10^{-6} \text{ mol KI} \times \frac{166.0 \text{ g}}{1 \text{ mol}} = 2.11 \times 10^{-4} \text{ g KI}$$

24.   Before any calculations are done, the equations must be *balanced*.

a. $Mg(s) + CuCl_2(aq) \rightarrow MgCl_2(aq) + Cu(s)$

molar mass of Mg = 24.31 g

$$25.0 \text{ g Mg} \times \frac{1 \text{ mol}}{24.31 \text{ g}} = 1.03 \text{ mol Mg}$$

$$1.03 \text{ mol Mg} \times \frac{1 \text{ mol } CuCl_2}{1 \text{ mol Mg}} = 1.03 \text{ mol } CuCl_2$$

b. $2AgNO_3(aq) + NiCl_2(aq) \rightarrow 2AgCl(s) + Ni(NO_3)_2(aq)$

molar mass of $AgNO_3$ = 169.9 g

$$25.0 \text{ g } AgNO_3 \times \frac{1 \text{ mol}}{169.9 \text{ g}} = 0.147 \text{ mol } AgNO_3$$

$$0.147 \text{ mol } AgNO_3 \times \frac{1 \text{ mol } NiCl_2}{2 \text{ mol } AgNO_3} = 0.0735 \text{ mol } NiCl_2$$

c. $NaHSO_3(aq) + NaOH(aq) \rightarrow Na_2SO_3(aq) + H_2O(l)$

molar mass of $NaHSO_3$ = 104.1 g

$$25.0 \text{ g } NaHSO_3 \times \frac{1 \text{ mol}}{104.1 \text{ g}} = 0.240 \text{ mol } NaHSO_3$$

$$0.240 \text{ mol } NaHSO_3 \times \frac{1 \text{ mol NaOH}}{1 \text{ mol } NaHSO_3} = 0.240 \text{ mol NaOH}$$

d. $KHCO_3(aq) + HCl(aq) \rightarrow KCl(aq) + H_2O(l) + CO_2(g)$

molar mass of $KHCO_3$ = 100.1 g

$$25.0 \text{ g } KHCO_3 \times \frac{1 \text{ mol}}{100.1 \text{ g}} = 0.250 \text{ mol } KHCO_3$$

$$0.250 \text{ mol } KHCO_3 \times \frac{1 \text{ mol HCl}}{1 \text{ mol } KHCO_3} = 0.250 \text{ mol HCl}$$

26.   a. $2BCl_3(s) + 3H_2(g) \rightarrow 2B(s) + 6HCl(g)$

molar masses: BCl_3, 117.16 g; B, 10.81 g; HCl, 36.46 g

$$15.0 \text{ g } BCl_3 \times \frac{1 \text{ mol } BCl_3}{117.16 \text{ g } BCl_3} = 0.128 \text{ mol } BCl_3$$

$$0.128 \text{ mol } BCl_3 \times \frac{2 \text{ mol B}}{2 \text{ mol } BCl_3} \times \frac{10.81 \text{ g B}}{1 \text{ mol B}} = 1.38 \text{ g B}$$

$$0.128 \text{ mol } BCl_3 \times \frac{6 \text{ mol HCl}}{2 \text{ mol } BCl_3} \times \frac{36.46 \text{ g HCl}}{1 \text{ mol HCl}} = 14.0 \text{ g HCl}$$

b. $2Cu_2S(s) + 3O_2(g) \rightarrow 2Cu_2O(s) + 2SO_2(g)$

molar masses: Cu_2S, 159.17 g; Cu_2O, 143.1 g; SO_2, 64.07 g

$$15.0 \text{ g } Cu_2S \times \frac{1 \text{ mol } Cu_2S}{159.17 \text{ g } Cu_2S} = 0.09424 \text{ mol } Cu_2S$$

$$0.09424 \text{ mol } Cu_2S \times \frac{2 \text{ mol } Cu_2O}{2 \text{ mol } Cu_2S} \times \frac{143.1 \text{ g } Cu_2O}{1 \text{ mol } Cu_2O} = 13.5 \text{ g } Cu_2O$$

$$0.09424 \text{ mol } Cu_2S \times \frac{2 \text{ mol } SO_2}{2 \text{ mol } Cu_2S} \times \frac{64.07 \text{ g } SO_2}{1 \text{ mol } SO_2} = 6.04 \text{ g } SO_2$$

c. $2Cu_2O(s) + Cu_2S(s) \rightarrow 6Cu(s) + SO_2(g)$

molar masses: Cu_2S, 159.17 g; Cu, 63.55 g; SO_2, 64.07 g

$$15.0 \text{ g } Cu_2S \times \frac{1 \text{ mol } Cu_2S}{159.17 \text{ g } Cu_2S} = 0.09424 \text{ mol } Cu_2S$$

$$0.09424 \text{ mol } Cu_2S \times \frac{6 \text{ mol Cu}}{1 \text{ mol } Cu_2S} \times \frac{63.55 \text{ g Cu}}{1 \text{ mol Cu}} = 35.9 \text{ g Cu}$$

$$0.09424 \text{ mol } Cu_2S \times \frac{1 \text{ mol } SO_2}{1 \text{ mol } Cu_2S} \times \frac{64.07 \text{ g } SO_2}{1 \text{ mol } SO_2} = 6.04 \text{ g } SO_2$$

d. $CaCO_3(s) + SiO_2(s) \rightarrow CaSiO_3(s) + CO_2(g)$

molar masses: SiO_2, 60.09 g; $CaSiO_3$, 116.17 g; CO_2, 44.01 g

$$15.0 \text{ g } SiO_2 \times \frac{1 \text{ mol } SiO_2}{60.09 \text{ g } SiO_2} = 0.2496 \text{ mol } SiO_2$$

$$0.2496 \text{ mol } SiO_2 \times \frac{1 \text{ mol } CaSiO_3}{1 \text{ mol } SiO_2} \times \frac{116.17 \text{ g } CaSiO_3}{1 \text{ mol } CaSiO_3} = 29.0 \text{ g } CaSiO_3$$

$$0.2496 \text{ mol } SiO_2 \times \frac{1 \text{ mol } CO_2}{1 \text{ mol } SiO_2} \times \frac{44.01 \text{ g } CO_2}{1 \text{ mol } CO_2} = 11.0 \text{ g } CO_2$$

28. $C_3H_8(l) + 5O_2(g) \rightarrow 3CO_2(g) + 4H_2O(g)$

molar masses: C_3H_8, 44.09 g; O_2 = 32.00 g

25.0 lb = 1.134×10^4 g

$$1.134 \times 10^4 \text{ g } C_3H_8 \times \frac{1 \text{ mol } C_3H_8}{44.09 \text{ g } C_3H_8} = 2.572 \times 10^2 \text{ mol } C_3H_8$$

$$2.572 \times 10^2 \text{ mol } C_3H_8 \times \frac{5 \text{ mol } O_2}{1 \text{ mol } C_3H_8} = 1.286 \times 10^3 \text{ mol } O_2$$

$$1.286 \times 10^3 \text{ mol } O_2 \times \frac{32.00 \text{ g } O_2}{1 \text{ mol } O_2} = 4.11 \times 10^4 \text{ g } O_2$$

30. $Cl_2(g) + 2KI(aq) \rightarrow I_2(s) + 2KCl(aq)$

molar masses: Cl_2, 70.90 g; I_2, 253.8 g

$$2.55 \text{ g } Cl_2 \times \frac{1 \text{ mol } Cl_2}{70.90 \text{ g } Cl_2} = 0.0360 \text{ mol } Cl_2$$

$$0.0360 \text{ mol } Cl_2 \times \frac{1 \text{ mol } I_2}{1 \text{ mol } Cl_2} = 0.0360 \text{ mol } I_2$$

$$0.0360 \text{ mol } I_2 \times \frac{253.8 \text{ g } I_2}{1 \text{ mol } I_2} = 9.13 \text{ g } I_2$$

32. $C_2H_5OH(l) + 3O_2(g) \rightarrow 2CO_2(g) + 3H_2O(l)$

molar masses: C_2H_5OH, 46.07 g; CO_2, 44.01 g

$$25.0 \text{ g } C_2H_5OH \times \frac{1 \text{ mol}}{46.07 \text{ g}} = 0.543 \text{ mol } C_2H_5OH$$

$$0.543 \text{ mol } C_2H_5OH \times \frac{2 \text{ mol } CO_2}{1 \text{ mol } C_2H_5OH} = 1.086 \text{ mol } CO_2$$

$$1.086 \text{ mol } CO_2 \times \frac{44.01 \text{ g}}{1 \text{ mol}} = 47.8 \text{ g } CO_2$$

34. $Co(OH)_2(s) \rightarrow CoO(s) + H_2O(g)$

molar masses: $Co(OH)_2$, 92.95 g; CoO, 74.93 g

$$5.75 \text{ g } Co(OH)_2 \times \frac{1 \text{ mol } Co(OH)_2}{92.95 \text{ g } Co(OH)_2} = 0.06186 \text{ mol } Co(OH)_2$$

$$0.06186 \text{ mol } Co(OH)_2 \times \frac{1 \text{ mol CoO}}{1 \text{ mol } Co(OH)_2} = 0.06186 \text{ mol CoO}$$

$$0.06186 \text{ mol CoO} \times \frac{74.93 \text{ g CoO}}{1 \text{ mol CoO}} = 4.64 \text{ g CoO}$$

36. $Cu(s) + S(s) \rightarrow CuS(s)$

molar masses: Cu, 63.55 g; S, 32.07 g

$$1.25 \text{ g Cu} \times \frac{1 \text{ mol}}{63.55 \text{ g}} = 1.97 \times 10^{-2} \text{ mol Cu}$$

$$1.97 \times 10^{-2} \text{ mol Cu} \times \frac{1 \text{ mol S}}{1 \text{ mol Cu}} = 1.97 \times 10^{-2} \text{ mol S}$$

$$1.97 \times 10^{-2} \text{ mol S} \times \frac{32.07 \text{ g}}{1 \text{ mol}} = 0.631 \text{ g S}$$

38. $2NaOH(aq) + CO_2(g) \rightarrow Na_2CO_3(aq) + H_2O(l)$

molar masses: NaOH, 40.00 g; CO_2, 44.01 g

$$5.00 \text{ g NaOH} \times \frac{1 \text{ mol}}{40.00 \text{ g}} = 0.125 \text{ mol NaOH}$$

$$0.125 \text{ mol NaOH} \times \frac{1 \text{ mol } CO_2}{2 \text{ mol NaOH}} = 6.25 \times 10^{-2} \text{ mol } CO_2$$

$$6.25 \times 10^{-2} \text{ mol } CO_2 \times \frac{44.01 \text{ g}}{1 \text{ mol}} = 2.75 \text{ g } CO_2$$

40. $2Mg(s) + O_2(g) \rightarrow 2MgO(s)$

molar masses: Mg, 24.31 g; MgO, 40.31 g

$$1.25 \text{ g Mg} \times \frac{1 \text{ mol}}{24.31 \text{ g}} = 5.14 \times 10^{-2} \text{ mol Mg}$$

$$5.14 \times 10^{-2} \text{ mol Mg} \times \frac{2 \text{ mol MgO}}{2 \text{ mol Mg}} = 5.14 \times 10^{-2} \text{ mol MgO}$$

$$5.14 \times 10^{-2} \text{ mol MgO} \times \frac{40.31 \text{ g}}{1 \text{ mol}} = 2.07 \text{ g MgO}$$

42. To determine the limiting reactant, first calculate the number of moles of each reactant present. Then determine how these numbers of moles correspond to the stoichiometric ratio indicated by the balanced chemical equation for the reaction.

44. A reactant is present *in excess* if there is more of that reactant present than is needed to combine with the limiting reactant for the process. By definition, the limiting reactant cannot be present in excess. An excess of any reactant does not affect the theoretical yield for a process: the theoretical yield is determined by the limiting reactant.

46. a. $S(s) + 2H_2SO_4(aq) \rightarrow 3SO_2(g) + 2H_2O(l)$

Molar masses: S, 32.07 g; H_2SO_4, 98.09 g; SO_2, 64.07 g; H_2O, 18.02 g

$$5.00 \text{ g S} \times \frac{1 \text{ mol}}{32.07 \text{ g}} = 0.1559 \text{ mol S}$$

$$5.00 \text{ g H}_2\text{SO}_4 \times \frac{1 \text{ mol}}{98.09 \text{ g}} = 0.05097 \text{ mol H}_2\text{SO}_4$$

According to the balanced chemical equation, we would need twice as much sulfuric acid as sulfur for complete reaction of both reactants. We clearly have much less sulfuric acid present than sulfur: sulfuric acid is the limiting reactant. The calculation of the masses of products produced is based on the number of moles of the sulfuric acid.

$$0.05097 \text{ mol H}_2\text{SO}_4 \times \frac{3 \text{ mol SO}_2}{2 \text{ mol H}_2\text{SO}_4} \times \frac{64.07 \text{ g SO}_2}{1 \text{ mol SO}_2} = 4.90 \text{ g SO}_2$$

$$0.05097 \text{ mol H}_2\text{SO}_4 \times \frac{2 \text{ mol H}_2\text{O}}{2 \text{ mol H}_2\text{SO}_4} \times \frac{18.02 \text{ g H}_2\text{O}}{1 \text{ mol H}_2\text{O}} = 0.918 \text{ mol H}_2\text{O}$$

b. $MnO_2(s) + 2H_2SO_4(aq) \rightarrow Mn(SO_4)_2 + 2H_2O(l)$

molar masses: MnO_2, 86.94 g; H_2SO_4 98.09 g;
$Mn(SO_4)_2$, 247.1 g; H_2O, 18.02 g

$$5.00 \text{ g MnO}_2 \times \frac{1 \text{ mol}}{86.94 \text{ g}} = 0.05751 \text{ mol MnO}_2$$

$$5.00 \text{ g H}_2\text{SO}_4 \times \frac{1 \text{ mol}}{98.09 \text{ g}} = 0.05097 \text{ mol H}_2\text{SO}_4$$

According to the balanced chemical equation, we would need twice as much sulfuric acid as manganese(IV) oxide for complete reaction of both reactants. We do not have this much sulfuric acid, so sulfuric acid must be the limiting reactant. The amount of each product produced will be based on the sulfuric acid reacting completely.

$$0.05097 \text{ mol H}_2\text{SO}_4 \times \frac{1 \text{ mol Mn(SO}_4)_2}{2 \text{ mol H}_2\text{SO}_4} \times \frac{247.1 \text{ g H}_2\text{SO}_4}{1 \text{ mol H}_2\text{SO}_4} = 6.30 \text{ g H}_2\text{SO}_4$$

$$0.05097 \text{ mol H}_2\text{SO}_4 \times \frac{2 \text{ mol H}_2\text{O}}{2 \text{ mol H}_2\text{SO}_4} \times \frac{18.02 \text{ g}}{1 \text{ mol}} = 0.918 \text{ mol H}_2\text{O}$$

c. $2H_2S(g) + 3O_2(g) \rightarrow 2SO_2(g) + 2H_2O(l)$

Molar masses: H_2S, 34.09 g; O_2, 32.00 g; SO_2, 64.07 g; H_2O, 18.02 g

$$5.00 \text{ g H}_2\text{S} \times \frac{1 \text{ mol}}{34.09 \text{ g}} = 0.1467 \text{ mol H}_2\text{S}$$

$$5.00 \text{ g O}_2 \times \frac{1 \text{ mol}}{32.00 \text{ g}} = 0.1563 \text{ mol O}_2$$

According to the balanced equation, we would need 1.5 times as much O_2 as H_2S for complete reaction of both reactants. We don't have that much O_2, so O_2 must be the limiting reactant taht will control the masses of each product produced.

$$0.1563 \text{ mol } O_2 \times \frac{2 \text{ mol } SO_2}{3 \text{ mol } O_2} \times \frac{64.07 \text{ g } SO_2}{1 \text{ mol } SO_2} = 6.67 \text{ g } SO_2$$

$$0.1563 \text{ mol } O_2 \times \frac{2 \text{ mol } H_2O}{3 \text{ mol } O_2} \times \frac{18.02 \text{ g } H_2O}{1 \text{ mol } H_2O} = 1.88 \text{ g } H_2O$$

d. $3AgNO_3(aq) + Al(s) \rightarrow 3Ag(s) + Al(NO_3)_3(aq)$

Molar masses: $AgNO_3$, 169.9 g; Al, 26.98 g;

Ag, 107.9 g; $Al(NO_3)_3$, 213.0 g

$$5.00 \text{ g } AgNO_3 \times \frac{1 \text{ mol}}{169.9 \text{ g}} = 0.02943 \text{ mol } AgNO_3$$

$$5.00 \text{ g Al} \times \frac{1 \text{ mol}}{26.98 \text{ g}} = 0.1853 \text{ mol Al}$$

According to the balanced chemical equation, we would need three moles of $AgNO_3$ for every mole of Al for complete reaction of both reactants. We in fact have fewer moles of $AgNO_3$ than aluminum, so $AgNO_3$ must be the limiting reactant. The amount of product produced is calculated from the number of moles of the limiting reactant present:

$$0.02943 \text{ mol } AgNO_3 \times \frac{3 \text{ mol } Ag}{3 \text{ mol } AgNO_3} \times \frac{107.9 \text{ g } Ag}{1 \text{ mol } Ag} = 3.18 \text{ g } Ag$$

$$0.02943 \text{ mol } AgNO_3 \times \frac{1 \text{ mol } Al(NO_3)_3}{3 \text{ mol } AgNO_3} \times \frac{213.0 \text{ g } Al(NO_3)_3}{1 \text{ mol } Al(NO_3)_3} = 2.09 \text{ g}$$

48.   a. $CS_2(l) + 3O_2(g) \rightarrow CO_2(g) + 2SO_2(g)$

Molar masses: CS_2, 76.15 g; O_2, 32.00 g; CO_2, 44.01 g

$$1.00 \text{ g } CS_2 \times \frac{1 \text{ mol}}{76.15 \text{ g}} = 0.01313 \text{ mol } CS_2$$

$$1.00 \text{ g } O_2 \times \frac{1 \text{ mol}}{32.00 \text{ g}} = 0.03125 \text{ mol } O_2$$

From the balanced chemical equation, we would need three times as much oxygen as carbon disulfide for complete reaction of both reactants. We do not have this much oxygen, and so oxygen must be the limiting reactant.

$$0.03125 \text{ mol } O_2 \times \frac{1 \text{ mol } CO_2}{3 \text{ mol } O_2} \times \frac{44.01 \text{ g}}{1 \text{ mol}} = 0.458 \text{ g } CO_2$$

b. $2NH_3(g) + CO_2(g) \rightarrow CN_2H_4O(s) + H_2O(l)$

Molar masses: NH_3, 17.03 g; CO_2, 44.01 g; H_2O, 18.02 g

$$1.00 \text{ g } NH_3 \times \frac{1 \text{ mol}}{17.03 \text{ g}} = 0.05872 \text{ mol } NH_3$$

$$1.00 \text{ g } CO_2 \times \frac{1 \text{ mol}}{44.01 \text{ g}} = 0.02272 \text{ mol } CO_2$$

The balanced chemical equation tells us that we would need twice as many moles of ammonia as carbon dioxide for complete reaction of both reactants. We have *more* than this amount of ammonia present, so the reaction will be limited by the amount of carbon dioxide present.

$$0.02272 \text{ mol } CO_2 \times \frac{1 \text{ mol } H_2O}{1 \text{ mol } CO_2} \times \frac{18.02 \text{ g}}{1 \text{ mol}} = 0.409 \text{ g } H_2O$$

c. $H_2(g) + MnO_2(s) \rightarrow MnO(s) + H_2O(l)$

Molar masses: H_2, 2.016 g; MnO_2, 86.94 g; H_2O, 18.02 g

$$1.00 \text{ g } H_2 \times \frac{1 \text{ mol}}{2.016 \text{ g}} = 0.496 \text{ mol } H_2$$

$$1.00 \text{ g } MnO_2 \times \frac{1 \text{ mol}}{86.94 \text{ g}} = 0.0115 \text{ mol } MnO_2$$

Because the coefficients of both reactants in the balanced chemical equation are the same, we would need equal amounts of both reactants for complete reaction. Therefore, manganese(IV) oxide must be the limiting reactant and controls the amount of product obtained.

$$0.0115 \text{ mol } MnO_2 \times \frac{1 \text{ mol } H_2O}{1 \text{ mol } MnO_2} \times \frac{18.02 \text{ g}}{1 \text{ mol}} = 0.207 \text{ g } H_2O$$

d. $I_2(s) + Cl_2(g) \rightarrow 2ICl(g)$

Molar masses: I_2, 253.8 g; Cl_2, 70.90 g; ICl, 162.35 g

$$1.00 \text{ g } I_2 \times \frac{1 \text{ mol}}{253.8 \text{ g}} = 0.00394 \text{ mol } I_2$$

$$1.00 \text{ g } Cl_2 \times \frac{1 \text{ mol}}{70.90 \text{ g}} = 0.0141 \text{ mol } Cl_2$$

From the balanced chemical equation, we would need equal amounts of I_2 and Cl_2 for complete reaction of both reactants. As we have much less iodine than chlorine, iodine must be the limiting reactant.

$$0.00394 \text{ mol } I_2 \times \frac{2 \text{ mol } ICl}{1 \text{ mol } I_2} \times \frac{162.35 \text{ g}}{1 \text{ mol}} = 1.28 \text{ g } ICl$$

As all the H_2SO_4 present would react with only a small portion of the Al present, H_2SO_4 is therefore the limiting reactant and will control the amount of product obtained.

$$0.510 \text{ mol } H_2SO_4 \times \frac{1 \text{ mol } Al_2(SO_4)_3}{3 \text{ mol } H_2SO_4} \times \frac{342.2 \text{ g}}{1 \text{ mol}} = 58.2 \text{ g } Al_2(SO_4)_3$$

50. a. $CO(g) + 2H_2(g) \rightarrow CH_3OH(l)$

CO is the limiting reactant; 11.4 mg CH_3OH

b. $2Al(s) + 3I_2(s) \rightarrow 2AlI_3(s)$

I_2 is the limiting reactant; 10.7 mg AlI_3

c. $Ca(OH)_2(aq) + 2HBr(aq) \rightarrow CaBr_2(aq) + 2H_2O(l)$

HBr is the limiting reactant; 12.4 mg $CaBr_2$; 2.23 mg H_2O

d. $2Cr(s) + 2H_3PO_4(aq) \rightarrow 2CrPO_4(s) + 3H_2(g)$

H_3PO_4 is the limiting reactant; 15.0 mg $CrPO_4$; 0.309 mg H_2

52. $2NH_3(g) + CO_2(g) \rightarrow CN_2H_4O(s) + H_2O(l)$

molar masses: NH_3, 17.03 g; CO_2, 44.01 g; CN_2H_4O, 60.06 g

$$100. \text{ g } NH_3 \times \frac{1 \text{ mol } NH_3}{17.03 \text{ g } NH_3} = 5.872 \text{ mol } NH_3$$

$$100. \text{ g } CO_2 \times \frac{1 \text{ mol } CO_2}{44.01 \text{ g } CO_2} = 2.272 \text{ mol } CO_2$$

See if CO_2 is the limiting reactant.

$$2.272 \text{ mol } CO_2 \times \frac{2 \text{ mol } NH_3}{1 \text{ mol } CO_2} = 4.544 \text{ mol } NH_3$$

CO_2 is indeed the limiting reactant.

$$2.272 \text{ mol } CO_2 \times \frac{1 \text{ mol } CN_2H_4O}{1 \text{ mol } CO_2} = 2.272 \text{ mol } CN_2H_4O$$

$$2.272 \text{ mol } CN_2H_4O \times \frac{60.06 \text{ g } CN_2H_4O}{1 \text{ mol } CN_2H_4O} = 136 \text{ g } CN_2H_4O$$

54. $4Fe(s) + 3O_2(g) \rightarrow 2Fe_2O_3(s)$

Molar masses: Fe, 55.85 g; Fe_2O_3, 159.7 g

$1.25 \text{ g Fe} \times \dfrac{1 \text{ mol}}{55.85 \text{ g}} = 0.0224 \text{ mol Fe present}$

Calculate how many mol of O_2 are required to react with this amount of Fe

$0.0224 \text{ mol Fe} \times \dfrac{3 \text{ mol } O_2}{4 \text{ mol Fe}} = 0.0168 \text{ mol } O_2$

Because we have more O_2 than this, Fe must be the limiting reactant.

$0.0224 \text{ mol Fe} \times \dfrac{2 \text{ mol } Fe_2O_3}{4 \text{ mol Fe}} = 0.0112 \text{ mol } Fe_2O_3$

$0.0112 \text{ mol } Fe_2O_3 \times \dfrac{159.7 \text{ g } Fe_2O_3}{1 \text{ mol } Fe_2O_3} = 1.79 \text{ g } Fe_2O_3$

56. $2CuSO_4(aq) + 5KI(aq) \rightarrow 2CuI(s) + KI_3(aq) + 2K_2SO_4(aq)$

molar masses: $CuSO_4$, 159.6 g; KI, 166.0 g; CuI, 190.5 g; KI_3, 419.8 g; K_2SO_4, 174.3 g

$0.525 \text{ g } CuSO_4 \times \dfrac{1 \text{ mol}}{159.6 \text{ g}} = 3.29 \times 10^{-3} \text{ mol } CuSO_4$

$2.00 \text{ g KI} \times \dfrac{1 \text{ mol}}{166.0 \text{ g}} = 0.0120 \text{ mol KI}$

To determine the limiting reactant, let's calculate what amount of KI would be needed to react with the given amount of $CuSO_4$ present.

$3.29 \times 10^{-3} \text{ mol } CuSO_4 \times \dfrac{5 \text{ mol KI}}{2 \text{ mol } CuSO_4} = 8.23 \times 10^{-3} \text{ mol KI}$

As we have more KI present than the amount required to react with the $CuSO_4$ present, $CuSO_4$ must be the limiting reactant, which will control the amount of products produced.

$3.29 \times 10^{-3} \text{ mol } CuSO_4 \times \dfrac{2 \text{ mol Cu I}}{2 \text{ mol } CuSO_4} \times \dfrac{190.5 \text{ g}}{1 \text{ mol}} = 0.627 \text{ g CuI}$

$3.29 \times 10^{-3} \text{ mol } CuSO_4 \times \dfrac{1 \text{ mol } KI_3}{2 \text{ mol } CuSO_4} \times \dfrac{419.8 \text{ g}}{1 \text{ mol}} = 0.691 \text{ g } KI_3$

$3.29 \times 10^{-3} \text{ mol } CuSO_4 \times \dfrac{2 \text{ mol } K_2SO_4}{2 \text{ mol } CuSO_4} \times \dfrac{174.3 \text{ g}}{1 \text{ mol}} = 0.573 \text{ g } K_2SO_4$

58. $SiO_2(s) + 3C(s) \rightarrow 2CO(g) + SiC(s)$

molar masses: SiO_2, 60.09 g; SiC, 40.10 g

$1.0 \text{ kg} = 1.0 \times 10^3 \text{ g}$

$$1.0 \times 10^3 \text{ g SiO}_2 \times \frac{1 \text{ mol SiO}_2}{60.09 \text{ g SiO}_2} = 16.64 \text{ mol SiO}_2$$

From the balanced chemical equation, if 16.64 mol of SiO_2 were to react completely (an excess of carbon is present), then 16.64 mol of SiC should be produced (the coefficients of SiO_2 and SiC are the same).

$$16.64 \text{ mol SiC} \times \frac{40.01 \text{ g SiC}}{1 \text{ mol SiC}} = 6.7 \times 10^2 \text{ g SiC} = 0.67 \text{ kg SiC}$$

60. If the reaction is performed in a solvent, the product may have a substantial solubility in the solvent; the reaction may come to equilibrium before the full yield of product is achieved (see Chapter 16); loss of product may occur through operator error.

62. $\text{Percent yield} = \dfrac{\text{actual yield}}{\text{theoretical yield}} \times 100 = \dfrac{1.279 \text{ g}}{1.352 \text{ g}} \times 100 = 94.60\%$

64. $2\text{LiOH}(s) + \text{CO}_2(g) \rightarrow \text{Li}_2\text{CO}_3(s) + \text{H}_2\text{O}(g)$

molar masses: LiOH, 23.95 g; CO_2, 44.01 g

$$155 \text{ g LiOH} \times \frac{1 \text{ mol LiOH}}{23.95 \text{ g LiOH}} \times \frac{1 \text{ mol CO}_2}{2 \text{ mol LiOH}} \times \frac{44.01 \text{ g CO}_2}{1 \text{ mol CO}_2} = 142 \text{ g CO}_2$$

As the cartridge has only absorbed 102 g CO_2 out of a total capacity of 142 g CO_2, the cartridge has absorbed

$$\frac{102 \text{ g}}{142 \text{ g}} \times 100 = 71.8\% \text{ of its capacity.}$$

66. $\text{Ba}^{2+}(aq) + \text{SO}_4^{2-}(aq) \rightarrow \text{BaSO}_4(s)$

molar masses: SO_4^{2-}, 96.07 g; $BaCl_2$, 208.2 g; $BaSO_4$, 233.4 g

$$1.12 \text{ g SO}_4^{2-} \times \frac{1 \text{ mol SO}_4^{2-}}{96.07 \text{ g SO}_4^{2-}} = 0.01166 \text{ mol SO}_4^{2-}$$

$$5.02 \text{ g BaCl}_2 \times \frac{1 \text{ mol BaCl}_2}{208.2 \text{ g BaCl}_2} = 0.02411 \text{ mol BaCl}_2 = 0.02411 \text{ mol Ba}^{2+}$$

SO_4^{2-} is the limiting reactant.

$$0.01166 \text{ mol SO}_4^{2-} \times \frac{1 \text{ mol BaSO}_4}{1 \text{ mol SO}_4^{2-}} \times \frac{233.4 \text{ g BaSO}_4}{1 \text{ mol BaSO}_4} = 2.72 \text{ g BaSO}_4$$

$\text{Percent yield} = \dfrac{\text{actual yield}}{\text{theoretical yield}} \times 100 = \dfrac{2.02 \text{ g}}{2.72 \text{ g}} \times 100 = 74.3\%$

68. $\text{NaCl}(aq) + \text{NH}_3(aq) + \text{H}_2\text{O}(l) + \text{CO}_2(s) \rightarrow \text{NH}_4\text{Cl}(aq) + \text{NaHCO}_3(s)$

molar masses: NH_3, 17.03 g; CO_2, 44.01 g; $NaHCO_3$, 84.01 g

$$10.0 \text{ g } NH_3 \times \frac{1 \text{ mol } NH_3}{17.03 \text{ g } NH_3} = 0.5872 \text{ mol } NH_3$$

$$15.0 \text{ g } CO_2 \times \frac{1 \text{ mol } CO_2}{44.01 \text{ g } CO_2} = 0.3408 \text{ mol } CO_2$$

CO_2 is the limiting reactant.

$$0.3408 \text{ mol } CO_2 \times \frac{1 \text{ mol } NaHCO_3}{1 \text{ mol } CO_2} = 0.3408 \text{ mol } NaHCO_3$$

$$0.3408 \text{ mol } NaHCO_3 \times \frac{84.01 \text{ g } NaHCO_3}{1 \text{ mol } NaHCO_3} = 28.6 \text{ g } NaHCO_3$$

70. $C_6H_{12}O_6(s) + 6O_2(g) \rightarrow 6CO_2(g) + 6H_2O(g)$

molar masses: glucose, 180.2 g; CO_2, 44.01 g

$$1.00 \text{ g glucose} \times \frac{1 \text{ mol glucose}}{180.2 \text{ g glucose}} = 5.549 \times 10^{-3} \text{ mol glucose}$$

$$5.549 \times 10^{-3} \text{ mol glucose} \times \frac{6 \text{ mol } CO_2}{1 \text{ mol glucose}} = 3.33 \times 10^{-2} \text{ mol } CO_2$$

$$3.33 \times 10^{-2} \text{ mol } CO_2 \times \frac{44.01 \text{ g } CO_2}{1 \text{ mol } CO_2} = 1.47 \text{ g } CO_2$$

72. $Ba^{2+}(aq) + SO_4^{2-}(aq) \rightarrow BaSO_4(s)$

millimolar ionic masses: Ba^{2+}, 137.3 mg; SO_4^{2-}, 96.07 mg; $BaCl_2$, 208.2 mg

$$150 \text{ mg } SO_4^{2-} \times \frac{1 \text{ mmol } SO_4^{2-}}{96.07 \text{ mg } SO_4^{2-}} = 1.56 \text{ millimol } SO_4^{2-}$$

As barium ion and sulfate ion react on a 1:1 stoichiometric basis, then 1.56 millimol of barium ion is needed, which corresponds to 1.56 millimol of $BaCl_2$

$$1.56 \text{ millimol } BaCl_2 \times \frac{208.2 \text{ mg } BaCl_2}{1 \text{ millimol } BaCl_2} = 325 \text{ milligrams } BaCl_2 \text{ needed}$$

74. a. $UO_2(s) + 4HF(aq) \rightarrow UF_4(aq) + 2H_2O(l)$

One molecule (formula unit) of uranium(IV) oxide will combine with four molecules of hydrofluoric acid, producing one uranium(IV) fluoride molecule and two water molecules. One mole of uranium(IV) oxide will combine with four moles of hydrofluoric acid to produce one mole of uranium(IV) fluoride and two moles of water.

b. $2NaC_2H_3O_2(aq) + H_2SO_4(aq) \rightarrow Na_2SO_4(aq) + 2HC_2H_3O_2(aq)$

Two molecules (formula units) of sodium acetate react exactly with one molecule of sulfuric acid, producing one molecule (formula unit) of sodium sulfate and two molecules of acetic acid. Two moles of sodium acetate will combine with one mole of sulfuric acid, producing one mole of sodium sulfate and two moles of acetic acid.

c. $Mg(s) + 2HCl(aq) \rightarrow MgCl_2(aq) + H_2(g)$

One magnesium atom will react with two hydrochloric acid molecules (formula units) to produce one molecule (formula unit) of magnesium chloride and one molecule of hydrogen gas. One mole of magnesium will combine with two moles of hydrochloric acid, producing one mole of magnesium chloride and one mole of gaseous hydrogen.

d. $B_2O_3(s) + 3H_2O(l) \rightarrow 2B(OH)_3(aq)$

One molecule of diboron trioxide will react exactly with three molecules of water, producing two molecules of boron trihydroxide (boric acid). One mole of diboron trioxide will combine with three moles of water to produce two moles of boron trihydroxide (boric acid).

76.   For O_2: $\dfrac{5 \text{ mol } O_2}{1 \text{ mol } C_3H_8}$ For CO_2: $\dfrac{3 \text{ mol } CO_2}{1 \text{ mol } C_3H_8}$ For H_2O: $\dfrac{4 \text{ mol } H_2O}{1 \text{ mol } C_3H_8}$

78.   a. $NH_3(g) + HCl(g) \rightarrow NH_4Cl(s)$

molar mass of NH_3 = 17.0 g

$$1.00 \text{ g } NH_3 \times \frac{1 \text{ mol } NH_3}{17.0 \text{ g } NH_3} = 0.0588 \text{ mol } NH_3$$

$$0.0588 \text{ mol } NH_3 \times \frac{1 \text{ mol } NH_4Cl}{1 \text{ mol } NH_3} = 0.0588 \text{ mol } NH_4Cl$$

b. $CaO(s) + CO_2(g) \rightarrow CaCO_3(s)$

molar mass CaO = 56.1 g

$$1.00 \text{ g CaO} \times \frac{1 \text{ mol CaO}}{56.1 \text{ g CaO}} = 0.0178 \text{ mol CaO}$$

$$0.0178 \text{ mol CaO} \times \frac{1 \text{ mol } CaCO_3}{1 \text{ mol CaO}} = 0.0178 \text{ mol } CaCO_3$$

c. $4Na(s) + O_2(g) \rightarrow 2Na_2O(s)$

molar mass Na = 22.99 g

$$1.00 \text{ g Na} \times \frac{1 \text{ mol Na}}{22.99 \text{ g Na}} = 0.0435 \text{ mol Na}$$

$$0.0435 \text{ mol Na} \times \frac{2 \text{ mol } Na_2O}{4 \text{ mol Na}} = 0.0217 \text{ mol } Na_2O$$

d. $2P(s) + 3Cl_2(g) \rightarrow 2PCl_3(l)$

molar mass P = 30.97 g

$$1.00 \text{ g P} \times \frac{1 \text{ mol P}}{30.97 \text{ g P}} = 0.0323 \text{ mol P}$$

$$0.0323 \text{ mol P} \times \frac{2 \text{ mol PCl}_3}{2 \text{ mol P}} = 0.0323 \text{ mol PCl}_3$$

80.   a.   molar mass HNO_3 = 63.0 g

$$5.0 \text{ mol HNO}_3 \times \frac{63.0 \text{ g HNO}_3}{1 \text{ mol HNO}_3} = 3.2 \times 10^2 \text{ g HNO}_3$$

b.   molar mass Hg = 200.6 g

$$0.000305 \text{ mol Hg} \times \frac{200.6 \text{ g Hg}}{1 \text{ mol Hg}} = 0.0612 \text{ g Hg}$$

c.   molar mass K_2CrO_4 = 194.2 g

$$2.31 \times 10^{-5} \text{ mol K}_2\text{CrO}_4 \times \frac{194.2 \text{ g K}_2\text{CrO}_4}{1 \text{ mol K}_2\text{CrO}_4} = 4.49 \times 10^{-3} \text{ g K}_2\text{CrO}_4$$

d.   molar mass $AlCl_3$ = 133.3 g

$$10.5 \text{ mol AlCl}_3 \times \frac{133.3 \text{ g AlCl}_3}{1 \text{ mol AlCl}_3} = 1.40 \times 10^3 \text{ g AlCl}_3$$

e.   molar mass SF_6 = 146.1 g

$$4.9 \times 10^4 \text{ mol SF}_6 \times \frac{146.1 \text{ g SF}_6}{1 \text{ mol SF}_6} = 7.2 \times 10^6 \text{ g SF}_6$$

f.   molar mass NH_3 = 17.0 g

$$125 \text{ mol NH}_3 \times \frac{17.0 \text{ g NH}_3}{1 \text{ mol NH}_3} = 2.13 \times 10^3 \text{ g NH}_3$$

g.   molar mass Na_2O_2 = 77.98 g

$$0.01205 \text{ mol Na}_2\text{O}_2 \times \frac{77.98 \text{ g Na}_2\text{O}_2}{1 \text{ mol Na}_2\text{O}_2} = 0.9397 \text{ g Na}_2\text{O}_2$$

82. $2SO_2(g) + O_2(g) \rightarrow 2SO_3(g)$

molar masses: SO_2, 64.07 g; SO_3, 80.07 g

150 kg = 1.5×10^5 g

$$1.5 \times 10^5 \text{ g SO}_2 \times \frac{1 \text{ mol SO}_2}{64.07 \text{ g SO}_2} = 2.34 \times 10^3 \text{ mol SO}_2$$

$$2.34 \times 10^3 \text{ mol SO}_2 \times \frac{2 \text{ mol SO}_3}{2 \text{ mol SO}_2} = 2.34 \times 10^3 \text{ mol SO}_3$$

$$2.34 \times 10^3 \text{ mol SO}_3 \times \frac{80.07 \text{ g SO}_3}{1 \text{ mol SO}_3} = 1.9 \times 10^5 \text{ g SO}_3 = 1.9 \times 10^2 \text{ kg SO}_3$$

84. $2Na_2O_2(s) + 2H_2O(l) \rightarrow 4NaOH(aq) + O_2(g)$

molar masses: Na_2O_2, 77.98 g; O_2, 32.00 g

$$3.25 \text{ g Na}_2O_2 \times \frac{1 \text{ mol Na}_2O_2}{77.98 \text{ g Na}_2O_2} = 0.0417 \text{ mol Na}_2O_2$$

$$0.0417 \text{ mol Na}_2O_2 \times \frac{1 \text{ mol O}_2}{2 \text{ mol Na}_2O_2} = 0.0209 \text{ mol O}_2$$

$$0.0209 \text{ mol O}_2 \times \frac{32.00 \text{ g O}_2}{1 \text{ mol O}_2} = 0.669 \text{ g O}_2$$

86. $Zn(s) + 2HCl(aq) \rightarrow ZnCl_2(aq) + H_2(g)$

molar masses: Zn, 65.38 g; H_2, 2.016 g

$$2.50 \text{ g Zn} \times \frac{1 \text{ mol Zn}}{65.38 \text{ g Zn}} = 0.03824 \text{ mol Zn}$$

$$0.03824 \text{ mol Zn} \times \frac{1 \text{ mol H}_2}{1 \text{ mol Zn}} = 0.03824 \text{ mol H}_2$$

$$0.03824 \text{ mol H}_2 \times \frac{2.016 \text{ g H}_2}{1 \text{ mol H}_2} = 0.0771 \text{ g H}_2$$

88.   a. $2Na(s) + Br_2(l) \rightarrow 2NaBr(s)$

molar masses: Na, 22.99 g; Br_2, 159.8 g; NaBr, 102.9 g

$$5.0 \text{ g Na} \times \frac{1 \text{ mol Na}}{22.99 \text{ g Na}} = 0.2175 \text{ mol Na}$$

$$5.0 \text{ g Br}_2 \times \frac{1 \text{ mol Br}_2}{159.8 \text{ g Br}_2} = 0.03129 \text{ mol Br}_2$$

Intuitively, we would suspect that Br_2 is the limiting reactant, because there is much less Br_2 than Na on a mole basis. To *prove* that Br_2 is the limiting reactant, the following calculation is needed:

$$0.03129 \text{ mol Br}_2 \times \frac{2 \text{ mol Na}}{1 \text{ mol Br}_2} = 0.06258 \text{ mol Na}.$$

Clearly, there is more Na than this present, so Br_2 limits the reaction extent and the amount of NaBr formed.

$$0.03129 \text{ mol Br}_2 \times \frac{2 \text{ mol NaBr}}{1 \text{ mol Br}_2} = 0.06258 \text{ mol NaBr}$$

$$0.06258 \text{ mol NaBr} \times \frac{102.9 \text{ g NaBr}}{1 \text{ mol NaBr}} = 6.4 \text{ g NaBr}$$

b. $Zn(s) + CuSO_4(aq) \rightarrow ZnSO_4(aq) + Cu(s)$

molar masses: Zn, 65.38 g; Cu, 63.55 g;

$$ZnSO_4, 161.5 \text{ g}; CuSO_4, 159.6 \text{ g}$$

$$5.0 \text{ g Zn} \times \frac{1 \text{ mol Zn}}{65.38 \text{ g Zn}} = 0.07648 \text{ mol Zn}$$

$$5.0 \text{ g CuSO}_4 \times \frac{1 \text{ mol CuSO}_4}{159.6 \text{ g CuSO}_4} = 0.03132 \text{ mol CuSO}_4$$

As the coefficients of Zn and $CuSO_4$ are the *same* in the balanced chemical equation, an equal number of moles of Zn and $CuSO_4$ would be needed for complete reaction. There is less $CuSO_4$ present, so $CuSO_4$ must clearly be the limiting reactant.

$$0.03132 \text{ mol CuSO}_4 \times \frac{1 \text{ mol ZnSO}_4}{1 \text{ mol CuSO}_4} = 0.03132 \text{ mol ZnSO}_4$$

$$0.03132 \text{ mol ZnSO}_4 \times \frac{161.5 \text{ g ZnSO}_4}{1 \text{ mol ZnSO}_4} = 5.1 \text{ g ZnSO}_4$$

$$0.03132 \text{ mol CuSO}_4 \times \frac{1 \text{ mol Cu}}{1 \text{ mol CuSO}_4} = 0.03132 \text{ mol Cu}$$

$$0.03132 \text{ mol Cu} \times \frac{63.55 \text{ g Cu}}{1 \text{ mol Cu}} = 2.0 \text{ g Cu}$$

c. $NH_4Cl(aq) + NaOH(aq) \rightarrow NH_3(g) + H_2O(l) + NaCl(aq)$

molar masses: NH_4Cl, 53.49 g; NaOH, 40.00 g; NH_3, 17.03 g

$$H_2O, 18.02 \text{ g}; NaCl, 58.44 \text{ g}$$

$$5.0 \text{ g NH}_4Cl \times \frac{1 \text{ mol NH}_4Cl}{53.49 \text{ g NH}_4Cl} = 0.09348 \text{ mol NH}_4Cl$$

$$5.0 \text{ g NaOH} \times \frac{1 \text{ mol NaOH}}{40.00 \text{ g NaOH}} = 0.1250 \text{ mol NaOH}$$

As the coefficients of NH_4Cl and NaOH are both *one* in the balanced chemical equation for the reaction, an equal number of moles of NH_4Cl and NaOH would be needed for complete reaction. There is less NH_4Cl present, so NH_4Cl must be the limiting reactant.
As the coefficients of the products in the balanced chemical equation are also all *one*, if 0.09348 mol of NH_4Cl (the limiting reactant) reacts completely, then 0.09348 mol of each product will be formed.

$$0.09348 \text{ mol NH}_3 \times \frac{17.03 \text{ g NH}_3}{1 \text{ mol NH}_3} = 1.6 \text{ g NH}_3$$

$$0.09348 \text{ mol H}_2\text{O} \times \frac{18.02 \text{ g H}_2\text{O}}{1 \text{ mol H}_2\text{O}} = 1.7 \text{ g H}_2\text{O}$$

$$0.09348 \text{ mol NaCl} \times \frac{58.44 \text{ g NaCl}}{1 \text{ mol NaCl}} = 5.5 \text{ g NaCl}$$

d. $Fe_2O_3(s) + 3CO(g) \rightarrow 2Fe(s) + 3CO_2(g)$

molar masses: Fe_2O_3, 159.7 g; CO, 28.01 g

Fe, 55.85 g; CO_2, 44.01 g

$$5.0 \text{ g Fe}_2\text{O}_3 \times \frac{1 \text{ mol Fe}_2\text{O}_3}{159.7 \text{ g Fe}_2\text{O}_3} = 0.03131 \text{ mol Fe}_2\text{O}_3$$

$$5.0 \text{ g CO} \times \frac{1 \text{ mol CO}}{28.01 \text{ g CO}} = 0.1785 \text{ mol CO}$$

Because there is considerably less Fe_2O_3 than CO on a mole basis, let's see if Fe_2O_3 is the limiting reactant.

$$0.03131 \text{ mol Fe}_2\text{O}_3 \times \frac{3 \text{ mol CO}}{1 \text{ mol Fe}_2\text{O}_3} = 0.09393 \text{ mol CO}$$

As there is 0.1785 mol of CO present, but we have determined that only 0.09393 mol CO would be needed to react with all the Fe_2O_3 present, then Fe_2O_3 must be the limiting reactant. CO is present in excess.

$$0.03131 \text{ mol Fe}_2\text{O}_3 \times \frac{2 \text{ mol Fe}}{1 \text{ mol Fe}_2\text{O}_3} = 0.06262 \text{ mol Fe}$$

$$0.06262 \text{ mol Fe} \times \frac{55.85 \text{ g Fe}}{1 \text{ mol Fe}} = 3.5 \text{ g Fe}$$

$$0.03131 \text{ mol Fe}_2\text{O}_3 \times \frac{3 \text{ mol CO}_2}{1 \text{ mol Fe}_2\text{O}_3} = 0.09393 \text{ mol CO}_2$$

$$0.09393 \text{ mol CO}_2 \times \frac{44.01 \text{ g CO}_2}{1 \text{ mol CO}_2} = 4.1 \text{ g CO}_2$$

90. $N_2H_4(l) + O_2(g) \rightarrow N_2(g) + 2H_2O(g)$

molar masses: N_2H_4, 32.05 g; O_2, 32.00 g; N_2, 28.02 g; H_2O, 18.02 g

$$20.0 \text{ g N}_2\text{H}_4 \times \frac{1 \text{ mol N}_2\text{H}_4}{32.05 \text{ g N}_2\text{H}_4} = 0.624 \text{ mol N}_2\text{H}_4$$

$$20.0 \text{ g O}_2 \times \frac{1 \text{ mol O}_2}{32.00 \text{ g O}_2} = 0.625 \text{ mol O}_2$$

The two reactants are present in nearly the required ratio for complete reaction (due to the 1:1 stoichiometry of the reaction and the very similar molar masses of the substances). We will consider N_2H_4 as the limiting reactant in the following calculations.

$$0.624 \text{ mol } N_2H_4 \times \frac{1 \text{ mol } N_2}{1 \text{ mol } N_2H_4} = 0.624 \text{ mol } N_2$$

$$0.624 \text{ mol } N_2 \times \frac{28.02 \text{ g } N_2}{1 \text{ mol } N_2} = 17.5 \text{ g } N_2$$

$$0.624 \text{ mol } N_2H_4 \times \frac{2 \text{ mol } H_2O}{1 \text{ mol } N_2H_4} = 1.248 \text{ mol } H_2O = 1.25 \text{ mol } H_2O$$

$$1.248 \text{ mol } H_2O \times \frac{18.02 \text{ g } H_2O}{1 \text{ mol } H_2O} = 22.5 \text{ g } H_2O$$

92. $12.5 \text{ g theory} \times \dfrac{40 \text{ g actual}}{100 \text{ g theory}} = 5.0 \text{ g}$

Cumulative Review: Chapters 8 and 9

2. On a microscopic basis, one mole of a substance represents Avogadro's number (6.022×10^{23}) of individual units (atoms or molecules) of the substance. On a macroscopic, more practical basis, one mole of a substance represents the amount of substance present when the molar mass of the substance in grams is taken (for example 12.01 g of carbon will be one mole of carbon). Chemists have chosen these definitions so that there will be a simple relationship between measurable amounts of substances (grams) and the actual number of atoms or molecules present, and so that the number of particles present in samples of *different* substances can easily be compared. For example, it is known that carbon and oxygen react by the reaction

$$C(s) \; + \; O_2(g) \; \rightarrow \; CO_2(g).$$

Chemists understand this equation to mean that one carbon atom reacts with one oxygen molecule to produce one molecule of carbon dioxide, and also that one mole (12.01 g) of carbon will react with one mole (32.00 g) of oxygen to produce one mole (44.01 g) of carbon dioxide.

4. The molar mass of a compound is the mass in grams of one mole of the compound (6.022×10^{23} molecules of the compound), and is calculated by summing the average atomic masses of all the atoms present in a molecule of the compound. For example, a molecule of the compound H_3PO_4 contains three hydrogen atoms, one phosphorus atom, and four oxygen atoms: the molar mass is obtained by adding up the average atomic masses of these atoms: molar mass H_3PO_4 = 3(1.008 g) + 1(30.97 g) + 4(16.00 g) = 97.99 g

6. The empirical formula of a compound represents the lowest ratio of the relative number of atoms of each type present in a molecule of the compound, whereas the molecular formula represents the actual number of atoms of each type present in a real molecule of the compound. For example, both acetylene (molecular formula C_2H_2) and benzene (molecular formula C_6H_6) have the same relative number of carbon and hydrogen atoms (one hydrogen for each carbon atom), and so have the same empirical formula (CH). Once the empirical formula of a compound has been determined, it is also necessary to determine the molar mass of the compound before the actual molecular formula can be calculated. As real molecules cannot contain fractional parts of atoms, the molecular formula is always a whole number multiple of the empirical formula. For the examples above, the molecular formula of acetylene is twice the empirical formula, and the molecular formula of benzene is six times the empirical formula (both factors are integers).

8. In question 7, we chose to calculate the percentage composition of phosphoric acid, H_3PO_4: 3.086% H, 31.60% P, 65.31% O. We could convert this percentage composition data into "experimental" data by first choosing a mass of sample to be "analyzed", and then calculating what mass of each element is present in this size sample using the percentage of each element. For example, suppose we choose our sample to have a mass

of 2.417 g. Then the masses of H, P, and O present in this sample would be given by the following:

$$g\ H\ =\ (2.417\ g\ sample)\ \times\ \frac{3.086\ g\ H}{100.0\ g\ sample}\ =\ 0.07459\ g\ H$$

$$g\ P\ =\ (2.417\ g\ sample)\ \times\ \frac{31.60\ g\ P}{100.0\ g\ sample}\ =\ 0.7638\ g\ P$$

$$g\ O\ =\ (2.417\ g\ sample)\ \times\ \frac{65.31\ g\ O}{100.0\ g\ sample}\ =\ 1.579\ g\ O$$

Note that (0.07459 g + 0.7638 g + 1.579 g) = 2.41739 = 2.417 g.

So our new problem could be worded as follows: "A 2.417 g sample of a compound has been analyzed and was found to contain 0.07459 g H, 0.7638 g of P, and 1.579 g of oxygen. Calculate the empirical formula of the compound".

$$mol\ H\ =\ (0.07459\ g\ H)\ \times\ \frac{1\ mol\ H}{1.008\ g\ H}\ =\ 0.07400\ mol\ H$$

$$mol\ P\ =\ (0.7638\ g\ P)\ \times\ \frac{1\ mol\ P}{30.97\ g\ P}\ =\ 0.02466\ mol\ P$$

$$mol\ O\ =\ (1.579\ g\ O)\ \times\ \frac{1\ mol\ O}{16.00\ g\ O}\ =\ 0.09869\ mol\ O$$

Dividing each of these numbers of moles by the smallest number of moles (0.02466 mol P) gives the following:

$$\frac{0.07400\ mol\ H}{0.02466}\ =\ 3.001\ mol\ H$$

$$\frac{0.02466\ mol\ P}{0.02466}\ =\ 1.000\ mol\ P$$

$$\frac{0.09869\ mol\ O}{0.02466}\ =\ 4.002\ mol\ O$$

The empirical formula is (not surprisingly) just H_3PO_4!

10.   The mole ratios for a reaction are based on the *coefficients* of the balanced chemical equation for the reaction: these coefficients show in what proportions molecules (or moles of molecules) combine. For a given amount of C_2H_5OH, the following mole ratios could be constructed, which would enable you to calculate the number of moles of each product, or of the second reactant, that would be involved.

$$C_2H_5OH(l)\ +\ 3O_2(g)\ \rightarrow\ 2CO_2(g)\ +\ 3H_2O(g)$$

to calculate mol CO_2 produced:

$$\frac{2 \text{ mol } CO_2}{1 \text{ mol } C_2H_5OH}$$

to calculate mol H_2O produced:

$$\frac{3 \text{ mol } H_2O}{1 \text{ mol } C_2H_5OH}$$

to calculate mol O_2 required:

$$\frac{3 \text{ mol } O_2}{1 \text{ mol } C_2H_5OH}$$

We could then calculate the numbers of moles of the other substances if 0.65 mol of C_2H_5OH were to be combusted as follows:

$$\text{mol } CO_2 \text{ produced} = (0.65 \text{ mol } C_2H_5OH) \times \frac{2 \text{ mol } CO_2}{1 \text{ mol } C_2H_5OH} = 1.3 \text{ mol } CO_2$$

$$\text{mol } H_2O \text{ produced} = (0.65 \text{ mol } C_2H_5OH) \times \frac{3 \text{ mol } H_2O}{1 \text{ mol } C_2H_5OH} = 1.95 = 2.0 \text{ mol } H_2O$$

$$\text{mol } O_2 \text{ required} = (0.65 \text{ mol } C_2H_5OH \times \frac{3 \text{ mol } O_2}{1 \text{ mol } C_2H_5OH} = 1.95 = 2.0 \text{ mol } O_2$$

12. Although we can calculate specifically the exact amounts of each reactant needed for a chemical reaction, oftentimes reaction mixtures are prepared using more or less arbitrary amounts of the reagents. However, regardless of how much of each reagent may be used for a reaction, the substances still react stoichiometrically, according to the mole ratios derived from the balanced chemical equation for the reaction. When arbitrary amounts of reactants are used, there will be one reactant which, stoichiometrically, is present in the least amount. This substance is called the *limiting reactant* for the experiment. It is the limiting reactant that controls how much product is formed, regardless of how much of the other reactants are present. The limiting reactant limits the amount of product that can form in the experiment, because once the limiting reactant has reacted completely, the reaction must stop. We say that the other reactants in the experiment are present in excess, which means that a portion of these reactants will still be present unchanged after the reaction has ended and the limiting reactant has been used up completely.

14. The *theoretical yield* for an experiment is the mass of product calculated based on the limiting reactant for the experiment being completely consumed. The *actual yield* for an experiment is the mass of product actually collected by the experimenter. Obviously, any experiment is restricted by the skills of the experimenter and by the inherent limitations of the experimental method being used. For these reasons, the actual yield is often *less* than the theoretical yield (most scientific writers report the actual or percentage yield for their experiments as an indication of the usefulness of their experiments). Although one would expect that the actual yield should never be more than the theoretical yield, in real experiments, sometimes this happens: however, an actual

yield greater than a theoretical yield is usually taken to mean that
something is *wrong* in either the experiment (for example, impurities may
be present, or the reaction may not occur as envisioned) or in the
calculations.

16. a.   element

b. $\% \text{ C} = \dfrac{12.01 \text{ g C}}{76.51 \text{ g}} \times 100 = 15.77\% \text{ C}$

c. $\% \text{ Al} = \dfrac{26.98 \text{ g Al}}{213.0 \text{ g}} \times 100 = 12.67\% \text{ Al}$

d. $\% \text{ C} = \dfrac{3(12.01 \text{ g C})}{44.09 \text{ g}} \times 100 = 81.72\% \text{ C}$

e.   element

f. $\% \text{ Ca} = \dfrac{40.08 \text{ g Ca}}{100.09 \text{ g}} \times 100 = 40.04\% \text{ Ca}$

g. $\% \text{ C} = \dfrac{12.01 \text{ g C}}{28.01 \text{ g}} \times 100 = 42.88\% \text{ C}$

h. $\% \text{ C} = \dfrac{12.01 \text{ g C}}{44.01 \text{ g}} \times 100 = 27.29\% \text{ C}$

i. $\% \text{ Ca} = \dfrac{40.08 \text{ g Ca}}{110.98 \text{ g}} \times 100 = 36.11\% \text{ Ca}$

j.   element

18. a. $2AgNO_3(aq) + CaSO_4(aq) \rightarrow Ag_2SO_4(s) + Ca(NO_3)_2(aq)$

Molar masses: $AgNO_3$, 169.9 g; Ag_2SO_4, 311.9 g; $Ca(NO_3)_2$, 164.1 g

$25.0 \text{ g } AgNO_3 \times \dfrac{1 \text{ mol}}{169.9 \text{ g}} = 0.147 \text{ mol } AgNO_3$

$0.147 \text{ mol } AgNO_3 \times \dfrac{1 \text{ mol } Ag_2SO_4}{2 \text{ mol } AgNO_3} = 0.0735 \text{ mol } Ag_2SO_4$

$0.0735 \text{ mol } Ag_2SO_4 \times \dfrac{311.9 \text{ g}}{1 \text{ mol}} = 22.9 \text{ g } Ag_2SO_4$

$0.147 \text{ mol } AgNO_3 \times \dfrac{1 \text{ mol } Ca(NO_3)_2}{2 \text{ mol } AgNO_3} = 0.0735 \text{ mol } Ca(NO_3)_2$

$0.0735 \text{ mol } Ca(NO_3)_2 \times \dfrac{164.1 \text{ g}}{1 \text{ mol}} = 12.1 \text{ g } Ca(NO_3)_2$

b. $2Al(s) + 6HNO_3(aq) \rightarrow 2Al(NO_3)_3(aq) + 3H_2(g)$

Molar masses: Al, 26.98 g; $Al(NO_3)_3$, 213.0 g; H_2, 2.016 g

$$25.0 \text{ g Al} \times \frac{1 \text{ mol}}{26.98 \text{ g}} = 0.927 \text{ mol Al}$$

$$0.927 \text{ mol Al} \times \frac{2 \text{ mol Al}(NO_3)_3}{2 \text{ mol Al}} = 0.927 \text{ mol Al}(NO_3)_3$$

$$0.927 \text{ mol Al}(NO_3)_3 \times \frac{213.0 \text{ g}}{1 \text{ mol}} = 197 \text{ g Al}(NO_3)_3$$

$$0.927 \text{ mol Al} \times \frac{3 \text{ mol } H_2}{2 \text{ mol Al}} = 1.39 \text{ mol } H_2$$

$$1.39 \text{ mol } H_2 \times \frac{2.016 \text{ g}}{1 \text{ mol}} = 2.80 \text{ g } H_2$$

c. $H_3PO_4(aq) + 3NaOH(aq) \rightarrow Na_3PO_4(aq) + 3H_2O(l)$

Molar masses: H_3PO_4, 97.99 g; Na_3PO_4, 163.9 g; H_2O, 18.02 g

$$25.0 \text{ g } H_3PO_4 \times \frac{1 \text{ mol}}{97.99 \text{ g}} = 0.255 \text{ mol } H_3PO_4$$

$$0.255 \text{ mol } H_3PO_4 \times \frac{1 \text{ mol } Na_3PO_4}{1 \text{ mol } H_3PO_4} = 0.255 \text{ mol } Na_3PO_4$$

$$0.255 \text{ mol } Na_3PO_4 \times \frac{163.9 \text{ g}}{1 \text{ mol}} = 41.8 \text{ g } Na_3PO_4$$

$$0.255 \text{ mol } H_3PO_4 \times \frac{3 \text{ mol } H_2O}{1 \text{ mol } H_3PO_4} = 0.765 \text{ mol } H_2O$$

$$0.765 \text{ mol } H_2O \times \frac{18.02 \text{ g}}{1 \text{ mol}} = 13.8 \text{ g } H_2O$$

d. $CaO(s) + 2HCl(aq) \rightarrow CaCl_2(aq) + H_2O(l)$

Molar masses: CaO, 56.08 g; $CaCl_2$, 111.0 g; H_2O, 18.02 g

$$25.0 \text{ g CaO} \times \frac{1 \text{ mol}}{56.08 \text{ g}} = 0.446 \text{ mol CaO}$$

$$0.446 \text{ mol CaO} \times \frac{1 \text{ mol } CaCl_2}{1 \text{ mol CaO}} = 0.446 \text{ mol } CaCl_2$$

$$0.446 \text{ mol } CaCl_2 \times \frac{111.0 \text{ g}}{1 \text{ mol}} = 49.5 \text{ g } CaCl_2$$

$$0.446 \text{ mol CaO} \times \frac{1 \text{ mol } H_2O}{1 \text{ mol CaO}} = 0.446 \text{ mol } H_2O$$

$$0.446 \text{ mol } H_2O \times \frac{18.02 \text{ g}}{1 \text{ mol}} = 8.04 \text{ g } H_2O$$

20.   molar masses: C, 12.01 g; CO, 28.01 g; CO_2, 44.01 g

$$5.00 \text{ g C} \times \frac{1 \text{ mol C}}{12.01 \text{ g C}} = 0.416 \text{ mol C}$$

CO: $0.416 \text{ mol C} \times \dfrac{2 \text{ mol CO}}{2 \text{ mol C}} \times \dfrac{28.01 \text{ g CO}}{1 \text{ mol CO}} = 11.7 \text{ g CO}$

CO_2 $0.416 \text{ mol C} \times \dfrac{1 \text{ mol } CO_2}{1 \text{ mol C}} \times \dfrac{44.01 \text{ g } CO_2}{1 \text{ mol } CO_2} = 18.3 \text{ g } CO_2$

2.   Rutherford's experiments did not indicate how the electrons of an atom are arranged or how they move through space. Rutherford suggested that the electrons might revolve around the nucleus like the planets revolve around the sun, but he could not explain why the negative electrons were not completely attracted into the positively-charged nucleus.

4.   The different forms of electromagnetic radiation are similar in that they all exhibit the same type of wave-like behavior and are propagated through space at the same speed (the speed of light). The types of electromagnetic radiation differ in their frequency (and wavelength) and in the resulting amount of energy carried per photon.

6.   The *speed* of electromagnetic radiation represents how fast a given wave moves through space. The *frequency* of electromagnetic radiation represents how many complete cycles of the wave pass a given point per second. For example, your favorite radio station broadcasts waves of a particular frequency that distinguishes it, but how long those waves take to reach you depends on their speed through space.

8.   "wave-particle nature" (duality)

10.   The energy of the emitted photon is *exactly the same* as the energy change within the atom.

12.   A photon having an energy corresponding to the energy difference between the two states is emitted by an atom in an excited state when it returns to its ground state.

14.   excited

16.   When excited hydrogen atoms emit their excess energy, the photons of radiation emitted are always of exactly the same wavelength and energy. We consider this to mean that the hydrogen atom possesses only certain allowed energy states, and that the photons emitted correspond to the atom changing from one of these allowed energy states to another of the allowed energy state. The energy of the photon emitted corresponds to the energy difference in the allowed states. If the hydrogen atom did not possess discrete energy levels, then we would expect the photons emitted to have random wavelengths and energies.

18.   The energy of an emitted photon is *identical* to the energy change within the atom that gave rise to the emitted photon.

20.   The ground state of an atom is its lowest possible energy state.

22.   According to Bohr, when an atom absorbs energy, the electron moves to an orbit farther from the nucleus; when an atom emits radiation, the electron moves to an orbit closer to the nucleus.

24.   Bohr's theory *explained* the experimentally *observed* line spectrum of hydrogen *exactly*. Bohr's theory was ultimately discarded because when attempts were made to extend the theory to atoms other than hydrogen,

the calculated properties did *not* correspond closely to experimental measurements.

26.   An orbit represents a definite, exact circular pathway around the nucleus in which an electron can be found. An orbital represents a region of space in which there is a high probability of finding the electron.

28.   Any experiment that sought to measure the exact location of an electron (such as shooting a beam of light at it) would cause the electron to move. Any measurement made would necessitate the application or removal of energy, which would disturb the electron from where it had been before the measurement.

30.   Pictures we draw to represent orbitals should only be interpreted as probability maps. They are not meant to represent that the electron moves only on the surface of, or within, the region drawn in the picture. As the mathematical probability of finding the electron never actually becomes zero on moving outward from the nucleus, scientists have decided that pictures of orbitals should represent a 90% probability that the electron will be found inside the region depicted in the drawing (for 100% probability, the orbital would have to encompass all space).

32.   The *p* orbitals, in general, have two lobes and are sometimes described as having a "dumbbell" shape. The 2*p* and 3*p* orbitals are similar in shape, and in fact there are three equivalent 2*p* or 3*p* orbitals in the 2*p* or 3*p* subshell. The orbitals differ in size, mean distance from the nucleus, and energy.

34.   *n* = 1

36.   The fourth principal energy level is divided into four sublevels: these sublevels are given the designations 4*s*, 4*p*, 4*d*, and 4*f*. The fifth principal energy level is divided into five sublevels: these are given the designations 5*s*, 5*p*, 5*d*, 5*f*, and 5*g*.

38.   The Pauli exclusion principle states that an orbital can hold a maximum of two electrons, and those two electrons must have opposite spins.

40.   increases; as you move out from the nucleus, there is more space and room for more sublevels.

42.   opposite

44.   a.   incorrect (the *n* = 1 shell has only the 1*s* subshell)

       b.   incorrect (the *n* = 2 shell has only 2*s* and 2*p* subshells)

       c.   correct (the *n* = 4 shell contains *s*, *p*, *d* and *f* subshells)

       d.   correct (the *n* = 5 shell contains *s*, *p*, *d*, *f*, and *g* subshells)

46. When a hydrogen atom is in its ground state, the electron is found in the 1*s* orbital. The 1*s* orbital has the lowest energy of all the possible hydrogen orbitals.

48. The elements in a given vertical column of the periodic table have the same valence electron configuration. Having the same valence electron configuration causes the elements in a given group to have similar chemical properties.

50. a. $1s^2\ 2s^2\ 2p^6\ 3s^2\ 3p^6\ 4s^2\ 3d^{10}\ 4p^6\ 5s^2$

    b. $1s^2\ 2s^2\ 2p^6\ 3s^2\ 3p^6\ 4s^2\ 3d^{10}$

    c. $1s^2$

    d. $1s^2\ 2s^2\ 2p^6\ 3s^2\ 3p^6\ 4s^2\ 3d^{10}\ 4p^5$

52. a. $1s^2\ 2s^2\ 2p^6\ 3s^2\ 3p^6\ 4s^1$

    b. $1s^2\ 2s^2\ 2p^6\ 3s^2\ 3p^5$

    c. $1s^2\ 2s^2\ 2p^5\ 3s^2$

    d. $1s^2\ 2s^2\ 2p^2$

54. a.   (↑↓) (↑↓) (↑↓)(↑↓)(↑↓) (↑↓) (↑)( )( )

              1*s*   2*s*   2*p*   3*s*   3*p*

    b.   (↑↓) (↑↓) (↑↓)(↑↓)(↑↓) (↑↓) (↑)(↑)(↑)

              1*s*   2*s*   2*p*   3*s*   3*p*

    c.   (↑↓) (↑↓) (↑↓)(↑↓)(↑↓) (↑↓) (↑↓)(↑↓)(↑↓) (↑↓)

              1*s*   2*s*   2*p*   3*s*   3*p*   4*s*

          (↑↓)(↑↓)(↑↓)(↑↓)(↑↓) (↑↓)(↑↓)(↑)

                      3*d*   4*p*

    d.   (↑↓) (↑↓) (↑↓)(↑↓)(↑↓) (↑↓) (↑↓)(↑↓)(↑↓)

              1*s*   2*s*   2*p*   3*s*   3*p*

56. For the representative elements (those filling *s* and *p* subshells), the group number gives the number of valence electrons.

    a.   one (6*s*)

    b.   three (2*s*, 2*p*)

    c.   eight (4*s*, 4*p*)

    d.   two (3*s*)

58.   The properties of Rb and Sr suggest that they are members of Groups 1 and 2, respectively, and so must be filling the 5*s* orbital. The 5*s* orbital is lower in energy (and fills before) the 4*d* orbitals.

60.   a.   [Ar] $4s^2$

       b.   [Rn] $7s^1$

       c.   [Kr] $5s^2\ 4d^1$

       d.   [Xe] $6s^2\ 4f^1\ 5d^1$

62.   a.   [Ne] $3s^2\ 3p^3$

       b.   [Ne] $3s^2\ 3p^5$

       c.   [Ne] $3s^2$

       d.   [Ar] $4s^2\ 3d^{10}$

64.   a.   one

       b.   two

       c.   zero

       d.   ten

66.   The *position* of the element (both in terms of vertical column and horizontal row) indicates which set of orbitals is being filled last.

       a.   5*f*   c.   4*f*

       b.   5*f*   d.   6*p*

68.   a.   [Kr] $5s^2\ 4d^8$

       b.   [Rn] $7s^2\ 5f^4\ 6d^1$

       c.   [Kr] $5s^1\ 4d^7$

       d.   [Xe] $6s^1\ 4f^{14}\ 5d^{10}$

70.   The metallic elements *lose* electrons and form *positive* ions (cations); the nonmetallic elements *gain* electrons and form *negative* ions (anions). Remember that the electron itself is *negatively* charged.

72.   All exist as *diatomic* molecules (F_2, Cl_2, Br_2, I_2); all are *non*metals; all have relatively high electronegativities; all form 1- ions in reacting with metallic elements.

74.   Elements at the *left* of a period (horizontal row) lose electrons more readily; at the left of a period (given principal energy level) the nuclear charge is the smallest and the electrons are least tightly held.

76. The elements of a given period (horizontal row) have valence electrons in the same principal energy level. Nuclear charge, however, increases across a period going from left to right. Atoms at the left side have smaller nuclear charges and hold onto their valence electrons less tightly.

78. The *nuclear charge* increases from left to right within a period, pulling progressively more tightly on the valence electrons.

80. Ionization energies decrease in going from top to bottom within a vertical group; ionization energies increase in going from left to right within a horizontal period.

    a.   Li   c.   Cl

    b.   Ca   d.   S

82. Atomic size increases in going from top to bottom within a vertical group; atomic size decreases in going from left to right within a horizontal period.

    a.   Na

    b.   S

    c.   N

    d.   F

84. speed of light

86. photons

88. quantized

90. orbital

92. transition metal

94. spins

96. a. $1s^2\ 2s^2\ 2p^6\ 3s^2\ 3p^6\ 4s^1$ [Ar] $4s^1$

     (↑↓) (↑↓) (↑↓)(↑↓)(↑↓) (↑↓) (↑↓)(↑↓)(↑↓) (↑)
      1s   2s   2p   3s   3p   4s

    b. $1s^2\ 2s^2\ 2p^6\ 3s^2\ 3p^6\ 4s^2\ 3d^2$ [Ar] $4s^2\ 3d^2$

     (↑↓) (↑↓) (↑↓)(↑↓)(↑↓) (↑↓) (↑↓)(↑↓)(↑↓) (↑↓) (↑)(↑)( )( )( )
      1s   2s   2p   3s   3p   4s   3d

c. $1s^2\ 2s^2\ 2p^6\ 3s^2\ 3p^2$ [Ne] $3s^2\ 3p^2$

(↑↓) (↑↓) (↑↓)(↑↓)(↑↓) (↑↓) (↑)(↑)( )
1s   2s   2p   3s   3p

d. $1s^2\ 2s^2\ 2p^6\ 3s^2\ 3p^6\ 4s^2\ 3d^6$ [Ar] $4s^2\ 3d^6$

(↑↓) (↑↓) (↑↓)(↑↓)(↑↓) (↑↓) (↑↓)(↑↓)(↑↓) (↑↓) (↑↓)(↑)(↑)(↑)(↑)
1s   2s   2p   3s   3p   4s   3d

e. $1s^2\ 2s^2\ 2p^6\ 3s^2\ 3p^6\ 4s^2\ 3d^{10}$ [Ar] $4s^2\ 3d^{10}$

(↑↓) (↑↓) (↑↓)(↑↓)(↑↓) (↑↓) (↑↓)(↑↓)(↑↓) (↑↓) (↑↓)(↑↓)(↑↓)(↑↓)(↑↓)
1s   2s   2p   3s   3p   4s   3d

98. a. ns^2 b. $ns^2\ np^5$

     c. $ns^2\ np^4$ d. ns^1

     e. $ns^2\ np^4$

100. a. $$\lambda = \frac{h}{mv}$$

$$\lambda = \frac{6.63 \times 10^{-34}\ \text{J s}}{(9.11 \times 10^{-31}\ \text{kg})[0.90(3.00 \times 10^8\ \text{m s}^{-1})]}$$

$\lambda = 2.7 \times 10^{-12}$ m (0.0027 nm)

b. 4.4×10^{-34} m

c. 2×10^{-35} m

The wavelengths for the ball and the person are *infinitesimally small*, whereas the wavelength for the electron is nearly the same order of magnitude as the diameter of a typical atom.

102. Light is emitted from the hydrogen atom only at certain fixed wavelengths. If the energy levels of hydrogen were *continuous*, a hydrogen atom would emit energy at all possible wavelengths.

104. The third principal energy level of hydrogen is divided into *three* sublevels (3s, 3p, and 3d); there is a *single* 3s orbital; there is a set of *three* 3p orbitals; there is a set of *five* 3d orbitals. See Figures 10.21-10.24 for the shapes of these orbitals.

106. a.   incorrect; the $n = 1$ shell has only the 1s subshell

     b.   correct

     c.   incorrect; the $n = 3$ shell has only 3s. 3p, and 3d subshells

d.   correct

e.   correct

f.   correct

108. a. $1s^2\ 2s^2\ 2p^6\ 3s^2\ 3p^6\ 4s^2\ 3d^{10}\ 4p^5$

b. $1s^2\ 2s^2\ 2p^6\ 3s^2\ 3p^6\ 4s^2\ 3d^{10}\ 4p^6\ 5s^2\ 4d^{10}\ 5p^6$

c. $1s^2\ 2s^2\ 2p^6\ 3s^2\ 3p^6\ 4s^2\ 3d^{10}\ 4p^6\ 5s^2\ 4d^{10}\ 5p^6\ 6s^2$

d. $1s^2\ 2s^2\ 2p^6\ 3s^2\ 3p^6\ 4s^2\ 3d^{10}\ 4p^4$

110. a.   five ($2s$, $2p$)   c.   one ($3s$)

b.   seven ($3s$, $3p$)   d.   three ($3s$, $3p$)

112. a.   [Kr] $5s^2\ 4d^2$ b.   [Kr] $5s^2\ 4d^{10}\ 5p^5$

c.   [Ar] $4s^2\ 3d^{10}\ 4p^2$ d.   [Xe] $6s^1$

114. The *position* of the element (both in terms of vertical column and horizontal row) indicates which set of orbitals is being filled last.

a. $3d$ b. $4d$

c. $5f$ d. $4p$

116. metals, low; nonmetals, high

118. Atomic size increases in going from top to bottom within a vertical group; atomic size decreases in going from left to right within a horizontal period.

a.   Ca

b.   P

c.   K

Chapter 11   Chemical Bonding

2.   The *bond energy* is the amount of energy required to *break* a chemical bond. The larger the bond energy, the stronger the chemical bond.

4.   Covalent bonding exists when two atoms mutually complete their valence shells by equal sharing of one or more electron pairs. H_2 is an example. Polar covalent bonding also involves the sharing of one or more electron pairs between two atoms, but the pairs are not shared equally by the two atoms. This typically happens when the two atoms are not identical and have different electronegativities. HF is an example.

6.   In H_2 and HF, the bonding is covalent in nature, with an electron pair being shared between the atoms. In H_2, the two atoms are identical (the sharing is equal); in HF, the two atoms are different (the sharing is unequal) and as a result the bond is polar. Both of these are in marked contrast to the situation in NaF: NaF is an ionic compound—an electron has been completely transferred from sodium to fluorine—producing separate ions.

8.   A bond is polar if the centers of positive and negative charge do not coincide at the same point. The bond has a negative end and a positive end. Polar bonds will exist in any molecule with nonidentical bonded atoms (although the molecule, as a whole, may not be polar if the bond dipoles cancel each other). Two simple examples are HF and HCl: in both cases, the negative center of charge is closer to the halogen atom.

10.   The level of polarity in a polar covalent bond is determined by the difference in electronegativity of the atoms in the bond.

12.   In each case, the element higher up within a given group on the periodic table, or to the right within a period, has the higher electronegativity.

   a.   K< Cr < Br

   b.   Ra < Ba < Ca

   c.   Na < Al < Cl

14.   Generally, covalent bonds between atoms of *different* elements are *polar*.

   a.   ionic

   b.   polar covalent

   c.   polar covalent

   d.   ionic

16.   For a bond to be polar covalent, the atoms involved in the bond must have different electronegativities (must be of different elements).

   a.   covalent (atoms of the same element)

   b.   covalent (atoms of the same element)

c.   polar covalent (different elements)

d.   polar covalent (different elements)

18.   The *degree* of polarity of a polar covalent bond is indicated by the mag-
      nitude of the difference in electronegativities of the elements involved:
      the larger the difference in electronegativity, the more polar the bond.
      Electronegativity differences are given in parentheses below:

      a.   H-O (1.4); H-N (0.9); the H-O bond is more polar.

      b.   H-N (0.9); H-F (1.9); the H-F bond is more polar.

      c.   H-O (1.4); H-F (1.9); the H-F bond is more polar.

      d.   H-O (1.4); H-Cl (0.9); the H-O bond is more polar.

20.   The greater the electronegativity difference between two atoms, the more
      ionic will be the bond between those two atoms.

      a.   Ca-Cl

      b.   K-F

      c.   Cu-F

      d.   Na-Br

22.   The presence of strong bond dipoles and a large overall dipole moment in
      water make it a polar substance overall. Among the properties of water
      dependent on its dipole moment are its freezing point, melting point,
      vapor pressure, and its ability to dissolve many substances.

24.   In a diatomic molecule containing two different elements, the more
      electronegative atom will be the negative end of the molecule, and the
      less electronegative atom will be the positive end.

      a.   H

      b.   Cl

      c.   I

26.   In the figures, the arrow points toward the more electronegative atom.

      a. $\delta+$ P→F $\delta-$

      b. $\delta+$ P→O $\delta-$

      c. $\delta+$ P→C $\delta-$

      d.   P and H have similar electronegativities.

28.   In the figures, the arrow points toward the more electronegative atom.

    a.   nonpolar

    b. $\delta+$ I→Br $\delta-$

    c. $\delta+$ I→F $\delta-$

    d. $\delta+$ At→I $\delta-$

30.   The metallic elements achieve an electron configuration analogous to the previous noble gas by losing electrons from their valence shells.

32.   Atoms in covalent molecules gain a configuration like that of a noble gas by sharing one or more pairs of electrons between atoms: such shared pairs of electrons "belong" to each of the atoms of the bond at the same time. In ionic bonding, one atom completely gives over one or more electrons to another atom, and the resulting ions behave independently of one another.

34.   a.   Cl $1s^2\ 2s^2\ 2p^6\ 3s^2\ 3p^5$

            Cl^- $1s^2\ 2s^2\ 2p^6\ 3s^2\ 3p^6$

        Ar has the same configuration as Cl^-.

    b.   O $1s^2\ 2s^2\ 2p^4$

            O^{2-} $1s^2\ 2s^2\ 2p^6$

        Ne has the same configuration as O^{2-}.

    c.   Al $1s^2\ 2s^2\ 2p^6\ 3s^2\ 3p^1$

            Al^{3+} $1s^2\ 2s^2\ 2p^6$

        Ne has the same configuration as Al^{3+}.

    d.   Ba $1s^2\ 2s^2\ 2p^6\ 3s^2\ 3p^6\ 4s^2\ 3d^{10}\ 4p^6\ 5s^2\ 4d^{10}\ 5p^6\ 6s^2$

            Ba^{2+} $1s^2\ 2s^2\ 2p^6\ 3s^2\ 3p^6\ 4s^2\ 3d^{10}\ 4p^6\ 5s^2\ 4d^{10}\ 5p^6$

        Xe has the same configuration as Ba^{2+}.

    e.   Se $1s^2\ 2s^2\ 2p^6\ 3s^2\ 3p^6\ 4s^2\ 3d^{10}\ 4p^4$

            Se^{2-} $1s^2\ 2s^2\ 2p^6\ 3s^2\ 3p^6\ 4s^2\ 3d^{10}\ 4p^6$

        Kr has the same configuration as Se^{2-}.

36.   a. Sr^{2+} (Sr has two electrons more than the noble gas Kr.)

    b. F^- (F has one electron fewer than the noble gas Ne.)

c. Li^+ (Li has one electron more than the noble gas He.)

d. O^{2-} (O has two electrons fewer than the noble gas Ne.)

38.   a. Na_2S Na has one electron more than a noble gas; S has two electrons fewer than a noble gas.

b.   BaSe   Ba has two electrons more than a noble gas; Se has two electrons fewer than a noble gas.

c. $MgBr_2$ Mg has two electrons more than a noble gas; Br has one electron less than a noble gas.

d. Li_3N Li has one electron more than a noble gas; N has three electrons fewer than a noble gas.

e.   KH   K has one electron more than a noble gas; H has one electron less than a noble gas.

40.   a. Ca^{2+}, [Ar]; N^{3-}, [Ne]

b. Mg^{2+}, [Ne]; S^{2-}, [Ar]

c. Al^{3+}, [Ne]; F^-, [Ne]

d. Ra^{2+}, [Rn]; Br^-, [Kr]

e. Cs^+, [Xe]; P^{3-}, [Ar]

42.   An ionic solid such as NaCl consists of an array of alternating positively- and negatively-charged ions: that is, each positive ion has as its nearest neighbors a group of negative ions, and each negative ion has a group of positive ions surrounding it. In most ionic solids, the ions are packed as tightly as possible.

44.   In forming an anion, an atom gains additional electrons in its outermost (valence) shell. Additional electrons in the valence shell increases the repulsive forces between electrons, so the outermost shell becomes larger to accommodate this.

46.   Relative ionic sizes are given in Figure 11.9. Within a given horizontal row of the periodic chart, negative ions tend to be larger than positive ions because the negative ions contain a larger number of electrons in the valence shell. Within a vertical group of the periodic table, ionic size increases from top to bottom. In general, positive ions are smaller than the atoms they come from, whereas negative ions are larger than the atoms they come from.

a. F^-

b. Cl^-

c.   Ca

d.   I⁻

48.   Relative ionic sizes are indicated in Figure 11.9. Within a given horizon-
       tal row of the periodic chart, negative ions tend to be larger than posi-
       tive ions because the negative ions contain a larger number of electrons
       in the valence shell. Within a vertical group of the periodic table,
       ionic size increases from top to bottom. In general, positive ions are
       smaller than the atoms they come from, whereas negative ions are larger
       than the atoms they come from.

a.   I⁻

b.   Cl⁻

c.   Cl⁻

d. S^{2-}

50.   When atoms form covalent bonds, they try to attain a valence electronic
       configuration similar to that of the following noble gas element. When
       the elements in the first few horizontal rows of the periodic table form
       covalent bonds, they will attempt to gain configurations similar to the
       noble gases helium (2 valence electrons, duet rule), and neon and argon
       (8 valence electrons, octet rule).

52.   These elements attain a total of eight valence electrons, making the
       valence electron configurations similar to those of the noble gases Ne
       and Ar.

54.   When two atoms in a molecule are connected by a triple bond, the atoms
       share three pairs of electrons (6 electrons) in completing their outer-
       most shells. A simple molecule containing a triple bond is acetylene,
       C_2H_2 H:C:::C:H.

56.   a.   Rb

       b.   :Cl:

       c.   :Kr:

       d.   Ba

       e.   ·P·

       f.   :At·

58.   a.   C provides 4; each Br provides 7; total valence electrons = 32

      b.   N provides 5; each O provides 6; total valence electrons = 17

      c.   each C provides 4; each H provides 1; total valence electrons = 30

      d.   each O provides 6; each H provides 1; total valence electrons = 14

60.   a. H_2 Each H provides 1 valence electron.
                         Total valence electrons = 2

                         H—H

      b.   HCl   H provides 1 valence electron.
                         Cl provides 7 valence electrons.
                         Total valence electrons = 8

                         H—C̈l̈:

      c. CF_4 C provides 4 valence electrons.
                         Each F provides 7 valence electrons.
                         Total valence electrons = 32

                         :F̈:
                         |
                   :F̈—C—F̈:
                         |
                         :F̈:

      d. C_2F_6 each C provides 4 valence electrons.
                         Each F provides 7 valence electrons.
                         Total valence electrons = 50

                   :F̈: :F̈:
                    | |
              :F̈—C—C—F̈:
                    | |
                   :F̈: :F̈:

62.   a. H_2S Each H provides 1 valence electron.
                         S provides 6 valence electrons.
                         Total valence electrons = 8

                         H—S̈—H

b. SiF_4 Si provides 4 valence electrons.
                    Each F provides 7 valence electrons.
                    Total valence electrons = 32

c. C_2H_4 Each C provides 4 valence electrons.
                    Each H provides 1 valence electron.
                    Total valence electrons = 12

d. C_3H_8 Each C provides 4 valence electrons.
                    Each H provides 1 valence electron.
                    Total valence electrons = 20

64.   a. NO_2 N provides 5 valence electrons.
                    Each O provides 6 valence electrons.
                    Total valence electrons = 17

Note that there are several Lewis structures possible because of the odd number of electrons (the unpaired electron can be located on different atoms in the Lewis structure).

b. H_2SO_4 Each H provides 1 valence electron.
                    S provides 6 valence electrons.
                    Each O provides 6 valence electrons.
                    Total valence electrons = 32

c. N_2O_4

Each N provides 5 valence electrons.
Each O provides 6 valence electrons.
Total valence electrons = 34

66.   a. ClO_3^-

Cl provides 7 valence electrons.
Each O provides 6 valence electrons.
The 1- charge means 1 additional electron.
Total valence electrons = 26

b. O_2^{2-}

Each O provides 6 valence electrons.
The 2- charge means two additional valence electrons.
Total valence electrons = 14

$$[:\ddot{O}\!-\!\ddot{O}:]^{2-}$$

c. $C_2H_3O_2^-$

Each C provides 4 valence electrons.
Each H provides 1 valence electron.
Each O provides 6 valence electrons.
The 1- charge means 1 additional valence electron.
Total valence electrons = 24

68.   a. HPO_4^{2-}

H provides 1 valence electron.
P provides 5 valence electrons.
Each O provides 6 valence electrons
The 2- charge means two additional valence electrons.
Total valence electrons = 32

b. $H_2PO_4^-$ Each H provides 1 valence electron.
                    P provides 5 valence electrons.
                    Each O provides 6 valence electrons.
                    The 1- charge means one additional valence electron.
                    Total valence electrons = 32

c. PO_4^{3-} P provides 5 valence electrons.
                    Each O provides 6 valence electrons.
                    The 3- charge means three additional valence electrons.
                    Total valence electrons = 32

70.   The geometric structure of NH_3 is that of a trigonal pyramid. The
       nitrogen atom of NH_3 is surrounded by four electron pairs (three are bond-
       ing, one is a lone pair). The H–N–H bond angle is somewhat less than
       109.5° (due to the presence of the lone pair).

72.   The geometric structure of SiF_4 is that of a tetrahedron. The silicon
       atom of SiF_4 is surrounded by four bonding electron pairs. The F–Si–F
       bond angle is the characteristic angle of the tetrahedron, 109.5°.

74.   The general molecular structure of a molecule is determined by (1) *how
       many electron pairs* surround the central atom in the molecule, and (2)
       which of those electron pairs are used for *bonding* to the other atoms of
       the molecule. Nonbonding electron pairs on the central atom do, however,
       cause minor changes in the bond angles, compared to the ideal regular
       geometric structure.

76.   You will remember from high school geometry, that two points in space are
       all that is needed to define a straight line. A diatomic molecule repre-
       sents two points (the nuclei of the atoms) in space.

78.   In NF_3, the nitrogen atom has *four* pairs of valence electrons, whereas in
       BF_3, there are only *three* pairs of valence electrons around the boron
       atom. The nonbonding electron pair on nitrogen in NF_3 pushes the three F
       atoms out of the plane of the N atom.

80.   If you draw the Lewis structures for these species, you will see that
       each of the indicated atoms is surrounded by *four pairs* of electrons with
       a *tetrahedral* orientation. Although the electron pairs are arranged

tetrahedrally, the overall geometric shape of the molecules may *not* be tetrahedral, depending on what atoms are *bonded* to the four pairs of electrons on the central atom.

82. a. tetrahedral (four electron pairs on C, and four atoms attached)

     b. nonlinear, *v*-shaped (four electron pairs on S, but only two atoms attached)

     c. tetrahedral (four electron pairs on Ge, and four atoms attached)

84. a. basically tetrahedral around the P atom (the hydrogen atoms are attached to two of the oxygen atoms and do not affect greatly the geometrical arrangement of the oxygen atoms around the phosphorus)

     b. tetrahedral (4 electron pairs on Cl, and 4 atoms attached)

     c. trigonal pyramidal (4 electron pairs on S, and 3 atoms attached)

86. a. approximately 109.5° (the molecule is *v*-shaped or nonlinear)

     b. approximately 109.5° (the molecule is trigonal pyramidal)

     c. 109.5°

     d. approximately 120° (the double bond makes the molecule flat)

88. Acetylene is a linear molecule with bond angles of 180°.

90. double

92. The bond with the larger electronegativity difference will be the more polar bond. See Figure 11.3 for electronegativities.

     a. S–F

     b. P–O

     c. C–H

94. The bond energy of a chemical bond is the quantity of energy required to break the bond and separate the atoms.

96. In each case, the element *higher up* within a group on the periodic table has the higher electronegativity.

     a. Be

     b. N

     c. F

98.   For a bond to be polar covalent, the atoms involved in the bond must have different electronegativities (must be of different elements).

    a.   polar covalent (different elements)

    b.   *non*polar covalent (two atoms of the same element)

    c.   polar covalent (different elements)

    d.   *non*polar covalent (atoms of the same element)

100.   In a diatomic molecule containing two different elements, the more electronegative atom will be the negative end of the molecule, and the *less* electronegative atom will be the positive end.

    a.   oxygen

    b.   bromine

    c.   iodine

102.   a.   Al $1s^2\ 2s^2\ 2p^6\ 3s^2\ 3p^1$
             $Al^{3+}\ 1s^2\ 2s^2\ 2p^6$
             Ne has the same configuration as Al^{3+}.

    b.   Br $1s^2\ 2s^2\ 2p^6\ 3s^2\ 3p^6\ 4s^2\ 3d^{10}\ 4p^5$

           $Br^-\ 1s^2\ 2s^2\ 2p^6\ 3s^2\ 3p^6\ 4s^2\ 3d^{10}\ 4p^6$

           Kr has the same configuration as Br^-.

    c.   Ca $1s^2\ 2s^2\ 2p^6\ 3s^2\ 3p^6\ 4s^2$

           $Ca^{2+}\ 1s^2\ 2s^2\ 2p^6\ 3s^2\ 3p^6$

           Ar has the same configuration as Ca^{2+}.

    d.   Li $1s^2\ 2s^1$

           $Li^+\ 1s^2$

           He has the same configuration as Li^+.

    e.   F $1s^2\ 2s^2\ 2p^5$

           $F^-\ 1s^2\ 2s^2\ 2p^6$

           Ne has the same configuration as F^-.

104.   a. Na_2Se Na has one electron more than a noble gas; Se has two electrons fewer than a noble gas.

    b.   RbF   Rb has one electron more than a noble gas; F has one electron less than a noble gas.

c. K₂Te   K has one electron more than a noble gas; Te has two electrons fewer than a noble gas.

d. BaSe   Ba has two electrons more than a noble gas; Se has two electrons fewer than a noble gas.

e. KAt   K has one electron more than a noble gas; At has one electron less than a noble gas.

f. FrCl   Fr has one electron more than a noble gas; Cl has one electron less than a noble gas.

106. Relative ionic sizes are indicated in Figure 11.9.

a. Na^+

b. Al^{3+}

c. F^-

d. Na^+

108. a. H provides 1; N provides 5; each O provides 6; total valence electrons = 24

b. each H provides 1; S provides 6; each O provides 6; total valence electrons = 32

c. each H provides 1; P provides 5; each O provides 6; total valence electrons = 32

d. H provides 1; Cl provides 7; each O provides 6; total valence electrons = 32

110. a. N_2H_4 Each N provides 5 valence electrons.
Each H provides 1 valence electron.
Total valence electrons = 14

b. C_2H_6 Each C provides 4 valence electrons.
Each H provides 1 valence electron.
Total valence electrons = 14

c. NCl_3 N provides 5 valence electrons.
Each Cl provides 7 valence electrons.
Total valence electrons = 26

d. $SiCl_4$ Si provides 4 valence electrons.
Each Cl provides 7 valence electrons.
Total valence electrons = 32

112. a. NO_3^- N provides 5 valence electrons.
Each O provides 6 valence electrons.
The 1- charge means one additional valence electron.
Total valence electrons = 24

b. CO_3^{2-} C provides 4 valence electrons.
Each O provides 6 valence electrons.
The 2- charge means two additional valence electrons.
Total valence electrons = 24

c. NH_4^+ N provides 5 valence electrons.
Each H provides 1 valence electron.
The 1+ charge means one less valence electron.
Total valence electrons = 8

114. a.   four electron pairs arranged tetrahedrally about C

     b.   four electron pairs arranged tetrahedrally about Ge

     c.   three electron pairs arranged trigonally (planar) around B

116. a. ClO_3^-, trigonal pyramid (lone pair on Cl)

     b. ClO_2^-, nonlinear (*v*-shaped, two lone pairs on Cl)

     c. ClO_4^-, tetrahedral (all pairs on Cl are bonding)

118. a.   nonlinear (*v*-shaped)

     b.   trigonal planar

     c.   basically trigonal planar around the C (the H is attached to one of the O atoms, and distorts the shape around the carbon only slightly)

     d.   linear

120. Ionic compounds tend to be hard, crystalline substances with relatively high melting and boiling points. Covalently bonded substances tend to be gases, liquids, or relatively soft solids, with much lower melting and boiling points.

Cumulative Review: Chapters 10 and 11

2.   An atom is said to be in its ground state when it is in its lowest
      possible energy state. When an atom possesses more energy than its
      ground state energy, the atom is said to be in an excited state. An atom
      is promoted from its ground state to an excited state by absorbing
      energy; when the atom returns from an excited state to its ground state
      it emits the excess energy as electromagnetic radiation. Atoms do not
      gain or emit radiation randomly, but rather do so only in discrete
      bundles of radiation called photons. The photons of radiation emitted by
      atoms are characterized by the wavelength (color) of the radiation:
      longer wavelength photons carry less energy than shorter wavelength
      photons. The energy of a photon emitted by an atom corresponds exactly
      to the difference in energy between two allowed energy states in an
      atom: thus, we can use an observable phenomenon (emission of light by
      excited atoms), to gain insight into the energy changes taking place
      within the atom.

4.   Bohr pictured the electron moving in only certain circular orbits around
      the nucleus. Each particular orbit (corresponding to a particular
      distance from the nucleus) had associated with it a particular energy
      (resulting from the attraction between the nucleus and the electron).
      When an atom absorbs energy, the electron moves from its ground state in
      the orbit closest to the nucleus ($n = 1$) to an orbit farther away from
      the nucleus ($n = 2, 3, 4, ...$). When an excited atom returns to its
      ground state, corresponding to the electron moving from an outer orbit
      to the orbit nearest the nucleus, the atom emits the excess energy as
      radiation. As the Bohr orbits are of fixed distances from the nucleus
      and from each other, when an electron moves from one fixed orbit to
      another, the energy change is of a definite amount. This corresponds to
      a photon being emitted of a particular characteristic wavelength and
      energy. The original Bohr theory worked very well for hydrogen: Bohr
      even predicted emission wavelengths for hydrogen that had not yet been
      seen, but were subsequently found at the exact wavelengths Bohr had
      calculated. However, when the simple Bohr model for the atom was applied
      to the emission spectra of other elements, the theory could not predict
      or explain the observed emission spectra.

6.   The lowest energy hydrogen atomic orbital is called the $1s$ orbital. The
      $1s$ orbital is spherical in shape (that is, the electron density around
      the nucleus is uniform in all directions from the nucleus). The $1s$
      orbital represents a probability map of electron density around the
      nucleus for the first principal energy level. The orbital does not have
      a sharp edge (it appears fuzzy) because the probability of finding the
      electron does not drop off suddenly with distance from the nucleus. The
      orbital does not represent just a spherical surface on which the
      electron moves (this would be similar to Bohr's original theory). When
      we draw a picture to represent the $1s$ orbital we are indicating that the
      probability of finding the electron within this region of space is
      greater than 90%. We know that the likelihood of finding the electron
      within this orbital is very high, but we still don't know exactly where
      in this region the electron is at a given instant in time.

8.   The third principal energy level of hydrogen is divided into three sublevels, the 3*s*, 3*p*, and 3*d* sublevels. The 3*s* subshell consists of the single 3*s* orbital: like the other *s* orbitals, the 3*s* orbital is spherical in shape. The 3*p* subshell consists of a set of three equal-energy 3*p* orbitals: each of these 3*p* orbitals has the same shape ("dumbbell"), but each of the 3*p* orbitals is oriented in a different direction in space. The 3*d* subshell consists of a set of five 3*d* orbitals: the 3*d* orbitals have the shapes indicated in Figure 10.23, and are oriented in different directions around the nucleus (students sometimes say that the 3*d* orbitals have the shape of a 4-leaf clover). The fourth principal energy level of hydrogen is divided into four sublevels, the 4*s*, 4*p*, 4*d*, and 4*f* orbitals. The 4*s* subshell consists of the single 4*s* orbital. The 4*p* subshell consists of a set of three 4*p* orbitals. The 4*d* subshell consists of a set of five 4*d* orbitals. The shapes of the 4*s*, 4*p*, and 4*d* orbitals are the same as the shapes of the orbitals of the third principal energy level (the orbitals of the fourth principal energy level are larger and further from the nucleus than the orbitals of the third level, however). The fourth principal energy level, because it is further from the nucleus, also contains a 4*f* subshell, consisting of seven 4*f* orbitals (the shapes of the 4*f* orbitals are beyond the scope of this text).

10.   Atoms have a series of principal energy levels symbolized by the letter *n*. The *n* = 1 level is the closest to the nucleus, and the energies of the levels increase as the value of *n* increases going out from the nucleus. Each principal energy level is divided into a set of sublevels of different characteristic shapes (designated by the letters *s, p, d,* and *f*). Each sublevel is further subdivided into a set of orbitals: each *s* subshell consists of a single *s* orbital; each *p* subshell consists of a set of three *p* orbitals; each *d* subshell consists of a set of five *d* orbitals; etc. A given orbital can be empty or it can contain one or two electrons, but never more than two electrons (if an orbital contains two electrons, then the electrons must have opposite intrinsic spins). The shape we picture for an orbital represents only a probability map for finding electrons: the shape does not represent a trajectory or pathway for electron movements.

12.   The valence electrons of atoms are the electrons in the outermost shell of the atom. The core electrons are those in principal energy levels closer to the nucleus than the outermost shell (the core electrons are the electrons that are not valence electrons). As the valence electrons are those in the outermost shell of the atom, it is these electrons that are affected by the presence of other atoms, and which are gained, lost, or shared with other atoms. The periodic table is basically arranged in terms of the valence electronic configurations of the elements (elements in the same vertical group have similar configurations): for example, Group 1 is the first column in the periodic table because all the elements in Group 1 have one valence electron.

14.   The general periodic table you drew for Question 13 should be similar to that found in Figure 10.27 of the text. From the column and row location of an element, you should be able to determine what the valence shell of

an element has for its electronic configuration. For example, the element in the third horizontal row, in the second vertical column, has $3s^2$ as its valence configuration. We know that the valence electrons are in the $n = 3$ shell because the element is in the third horizontal row. We know that the valence electrons are s electrons because the first two electrons in a horizontal row are always in an s subshell. We know that there are two electrons because the element is the second element in the horizontal row. As an additional example, the element in the seventh vertical column of the second horizontal row in the periodic table has valence configuration $2s^2 2p^5$.

16.   The ionization energy of an atom represents the energy required to remove an electron from the atom. As one goes from top to bottom in a vertical group in the periodic table, the ionization energies decrease (it becomes easier to remove an electron). As one goes down within a group, the valence electrons are farther and farther from the nucleus and are less tightly held. The ionization energies increase when going from left to right within a horizontal row within the periodic table. The left-hand side of the periodic table is where the metallic elements are found, which lose electrons relatively easily. The right-hand side of the periodic table is where the nonmetallic elements are found: rather than losing electrons, these elements tend to gain electrons. Within a given horizontal row in the periodic table, the valence electrons are all in the same principal energy shell: however, as you go from left to right in the horizontal row, the nuclear charge that holds onto the electrons is increasing one unit with each successive element, making it that much more difficult to remove an electron. The relative sizes of atoms also vary systematically with the location of an element in the periodic table. Within a given vertical group, the atoms get progressively larger when going from the top of the group to the bottom: the valence electrons of the atoms are in progressively higher principal energy shells (and are progressively further from the nucleus) as we go down in a group. In going from left to right within a horizontal row in the periodic table, the atoms get progressively smaller. Although all the elements in a given horizontal row in the periodic table have their valence electrons in the same principal energy shell, the nuclear charge is progressively increasing from left to right, making the given valence shell progressively smaller as the electrons are drawn more closely to the nucleus.

18.   Ionic bonding results when elements of very different electro-negativities react with each other. Typically a metallic element reacts with a nonmetallic element; the metallic element losing electrons and forming forming positive ions and the nonmetallic element gaining electrons and forming negative ions. Sodium chloride, $NaCl$, is an example of a typical ionic compound. The aggregate form of such a compound consists of a crystal lattice of alternating positively and negatively charged ions. A given positive ion is attracted by several surrounding negatively charged ions, and a given negative ion is attracted by several surrounding positively charged ions. Similar electrostatic attractions go on in three dimensions throughout the crystal of ionic solid, leading to a very stable system (with very high

melting and boiling points, for example). We know that ionically bonded
solids do not conduct electricity in the solid state (because the ions
are held tightly in place by all the attractive forces), but such
substances are strong electrolytes when melted or when dissolved in
water (either process sets the ions free to move around).

20. Electronegativity represents the relative ability of an atom in a
molecule to attract shared electrons towards itself. In order for a bond
to be polar, one of the atoms in the bond must attract the shared
electron pair towards itself and away from the other atom of the bond:
this can only happen if one atom of the bond is more electronegative
than the other (that is, that there is a considerable difference in
electronegativity for the two atoms of the bond). If two atoms in a bond
have the same electronegativity, then the two atoms pull the electron
pair equally and the bond is nonpolar covalent. If two atoms sharing a
pair of electrons have vastly different electronegativities, the
electron pair will be pulled so strongly by the more electronegative
atom that a negative ion may be formed (as well as a positive ion for
the second atom) and ionic bonding will result. If the difference in
electronegativity between two atoms sharing an electron pair is
somewhere in between these two extremes (equal sharing of the electron
pair and formation of ions), then a polar covalent bond results.

22. It has been observed over many experiments that when an active metal
like sodium or magnesium reacts with a nonmetal, the sodium atoms always
form Na^+ ions and the magnesium atoms always form Mg^{2+} ions. It has been
further observed that aluminum always forms only the Al^{3+} ion. When
nitrogen, oxygen, or fluorine form simple ions, the ions that are formed
are always N^{3-}, O^{2-}, and F^-, respectively. Clearly the facts that these
elements always form the same ions and that those ions all contain eight
electrons in the outermost shell, led scientists to speculate that there
must be something fundamentally stable about a species that has eight
electrons in its outermost shell (like the noble gas neon). The repeated
observation that so many elements, when reacting, tend to attain an
electronic configuration that is isoelectronic with a noble gas led
chemists to speculate that all elements try to attain such a
configuration for their outermost shells. In general, when atoms of a
metal react with atoms of a nonmetal, the metal atoms lose electrons
until they have the configuration of the preceding noble gas, and the
nonmetal atoms gain electrons until they have the configuration of the
following noble gas. Covalently and polar covalently bonded molecules
also strive to attain pseudo-noble gas electronic configurations. For a
covalently bonded molecule like F_2, in which neither fluorine atom has a
greater tendency than the other to gain or lose electrons completely,
each F atom provides one electron of the pair of electrons that
constitutes the covalent bond. Each F atom feels also the influence of
the other F atom's electron in the shared pair, and each F atom
effectively fills its outermost shell. Similarly, in polar covalently
bonded molecules like HF or HCl, the shared pair of electrons between
the atoms effectively completes the outer electron shell of each atom
simultaneously to give each atom a noble gas-like electronic
configuration.

24.   Bonding between atoms to form a molecule involves only the valence
       electrons of the atoms (not the inner core electrons). So when we draw
       the Lewis structure of a molecule, we show only these valence electrons
       (both bonding valence electrons and nonbonding valence electrons,
       however). The most important requisite for the formation of a stable
       compound (which we try to demonstrate when we write Lewis structures) is
       that each atom of a molecule attains a noble gas electron configuration.
       When we write Lewis structures, we arrange the bonding and nonbonding
       valence electrons to try to complete the octet (or duet) for as many
       atoms as is possible.

26.   Obviously, you could choose practically any molecule for your
       discussion. Let's illustrate the method for ammonia, NH_3. First count up
       the total number of valence electrons available in the molecule (without
       regard to what atom they officially come from); remember that for the
       representative elements, the number of valence electrons is indicated by
       what group the element is found in on the periodic table. For NH_3,
       because nitrogen is in Group 5, one nitrogen atom would contribute five
       valence electrons. Because hydrogen atoms only have one electron each,
       the three hydrogen atoms provide an additional three valence electrons,
       for a total of eight valence electrons overall. Next write down the
       symbols for the atoms in the molecule, and use one pair of electrons
       (represented by a line) to form a bond between each pair of bound atoms.

$$H\!-\!N\!-\!H$$
$$|$$
$$H$$

       These three bonds use six of the eight valence electrons. Since each
       hydrogen already has its duet in what we have drawn so far, while the
       nitrogen atom only has six electrons around it so far, the final two
       valence electrons must represent a lone pair on the nitrogen.

$$H\!-\!\ddot{N}\!-\!H$$
$$|$$
$$H$$

28.   There are several types of exceptions to the octet rule described in the
       text. The octet rule is really a "rule of thumb" which we apply to
       molecules unless we have some evidence that a molecule does not follow
       the rule. There are some common molecules that, from experimental
       measurements, we know do not follow the octet rule. Boron and beryllium
       compounds sometimes do not fit the octet rule. For example, in BF_3, the
       boron atom only has six valence electrons, whereas in BeF_2, the
       beryllium atom only has four valence electrons. Other molecules that are
       exceptions to the octet rule include any molecule with an odd number of
       valence electrons (such as NO or NO_2): you can't get an octet (an even
       number) of electrons around each atom in a molecule with an odd number

of valence electrons. Even the oxygen gas we breathe is an exception to the octet rule: although we can write a Lewis structure for O_2 satisfying the octet rule for each oxygen, we know from experiment that O_2 contains unpaired electrons (which would not be consistent with a structure in which all the electrons were paired up.)

30.

Number of Valence Pairs	Bond Angle	Example(s)
2	180°	BeF_2, BeH_2
3	120°	BCl_3
4	109.5°	CH_4, CCl_4, GeF_4

32.   a.   Mg $1s^2 2s^2 2p^6 3s^2$
            forms Mg^{2+} ion

     b. Na^+ $1s^2 2s^2 2p^6$
            isoelectronic with [Ne]

     c.   Cl $1s^2 2s^2 2p^6 3s^2 3p^5$
            forms Cl^- ion

     d. F^- $1s^2 2s^2 2p^6$
            isoelectronic with [Ne]

     e.   Ba $1s^2 2s^2 2p^6 3s^2 3p^6 4s^2 3d^{10} 4p^6 5s^2 4d^{10} 5p^6 6s^2$
            forms Ba^{2+} ion

     f. I^- $1s^2 2s^2 2p^6 3s^2 3p^6 4s^2 3d^{10} 4p^6 5s^2 4d^{10} 5p^6$
            isoelectronic with [Xe]

     g.   Zn $1s^2 2s^2 2p^6 3s^2 3p^6 4s^2 3d^{10}$
            forms Zn^{2+} ion

     h. O^{2-} $1s^2 2s^2 2p^6$
            isoelectronic with [Ne]

     i. Al^{3+} $1s^2 2s^2 2p^6$
            isoelectronic with [Ne]

     j. P^{3-} $1s^2 2s^2 2p^6 3s^2 3p^6$
            isoelectronic with [Ar]

Chapter 12   Gases

2. Solids and liquids have essentially fixed volumes and are not able to be compressed easily. Gases have volumes that depend on their conditions, and can be compressed or expanded by changes in those conditions. Although the particles of matter in solids are essentially fixed in position (the solid is rigid), the particles in liquids and gases are free to move.

4. Figure 12.2 in the text shows a simple mercury barometer: a tube filled with mercury is inverted over a reservoir containing mercury that is open to the atmosphere. When the tube is inverted, the mercury falls to a level at which the pressure of the atmosphere is sufficient to support the column of mercury. One standard atmosphere of pressure is taken to be the pressure capable of supporting a column of mercury to a height of 760.0 mm above the reservoir level.

6. 760 (defined quantity)

8. a. $105.2 \text{ kPa} \times \dfrac{1 \text{ atm}}{101.325 \text{ kPa}} = 1.038 \text{ atm}$

    b. $75.2 \text{ cm Hg} \times \dfrac{10 \text{ mm Hg}}{1 \text{ cm Hg}} \times \dfrac{1 \text{ atm}}{760 \text{ mm Hg}} = 0.989 \text{ atm}$

    c. $752 \text{ mm Hg} \times \dfrac{1 \text{ atm}}{760 \text{ mm Hg}} = 0.989 \text{ atm}$

    d. $767 \text{ torr} = 767 \text{ mm Hg}$

        $767 \text{ torr} \times \dfrac{1 \text{ atm}}{760 \text{ torr}} = 1.01 \text{ atm}$

10. a. $0.9975 \text{ atm} \times \dfrac{760 \text{ mm Hg}}{1 \text{ atm}} = 758.1 \text{ mm Hg}$

    b. $225{,}400 \text{ Pa} \times \dfrac{760 \text{ mm Hg}}{101{,}325 \text{ Pa}} = 1691 \text{ mm Hg}$

    c. $99.7 \text{ kPa} \times \dfrac{760 \text{ mm Hg}}{101.325 \text{ kPa}} = 748 \text{ mm Hg}$

    d. $1.078 \text{ atm} \times \dfrac{760 \text{ mm Hg}}{1 \text{ atm}} = 819.3 \text{ mm Hg}$

12. a. $2.07 \times 10^6 \text{ Pa} \times \dfrac{1 \text{ kPa}}{10^3 \text{ Pa}} = 2.07 \times 10^3 \text{ kPa}$

    b. $795 \text{ mm Hg} \times \dfrac{101{,}325 \text{ Pa}}{760 \text{ mm Hg}} \times \dfrac{1 \text{ kPa}}{10^3 \text{ Pa}} = 106 \text{ kPa}$

c. $10.9 \text{ atm} \times \dfrac{101,325 \text{ Pa}}{1 \text{ atm}} \times \dfrac{1 \text{ kPa}}{10^3 \text{ Pa}} = 1.10 \times 10^3 \text{ kPa}$

d. $659 \text{ torr} \times \dfrac{101,325 \text{ Pa}}{760 \text{ torr}} \times \dfrac{1 \text{ kPa}}{10^3 \text{ Pa}} = 87.9 \text{ kPa}$

14.   increases

16. $PV = k;\qquad P_1V_1 = P_2V_2$

18.   a. $P_1 = 1.00 \text{ atm}$ $\qquad\qquad\qquad$ $P_2 = 699 \text{ torr} = 0.920 \text{ atm}$

$\qquad\qquad$ $V_1 = 541 \text{ mL}$ $\qquad\qquad\qquad\quad$ $V_2 = ? \text{ mL}$

$\qquad\qquad$ $V_2 = \dfrac{P_1V_1}{P_2} = \dfrac{(1.00 \text{ atm})(541 \text{ mL})}{(0.920 \text{ atm})} = 588 \text{ mL}$

b. $P_1 = 110.2 \text{ kPa}$ $\qquad\qquad\qquad$ $P_2 = 0.995 \text{ atm} = 100.8 \text{ kPa}$

$\qquad\qquad$ $V_1 = 2.32 \text{ L}$ $\qquad\qquad\qquad\quad$ $V_2 = ? \text{ L}$

$\qquad\qquad$ $V_2 = \dfrac{P_1V_1}{P_2} = \dfrac{(110.2 \text{ kPa})(2.32 \text{ L})}{(100.8 \text{ kPa})} = 2.54 \text{ L}$

c. $P_1 = 135 \text{ atm} = 1.026 \times 10^5 \text{ mm Hg}$ $\qquad$ $P_2 = ? \text{ mm Hg}$

$\qquad\qquad$ $V_1 = 4.15 \text{ mL}$ $\qquad\qquad\qquad\qquad\qquad\qquad$ $V_2 = 10.0 \text{ mL}$

$\qquad\qquad$ $P_2 = \dfrac{P_1V_1}{V_2} = \dfrac{(1.026 \times 10^5 \text{ mm Hg})(4.15 \text{ mL})}{(10.0 \text{ mL})} = 4.26 \times 10^4 \text{ mm Hg}$

20.   a. $P_1 = 1.07 \text{ atm}$ $\qquad\qquad\qquad$ $P_2 = 2.14 \text{ atm}$

$\qquad\qquad$ $V_1 = 291 \text{ mL}$ $\qquad\qquad\qquad\quad$ $V_2 = ? \text{ mL}$

$\qquad\qquad$ $V_2 = \dfrac{P_1V_1}{P_2} = \dfrac{(1.07 \text{ atm})(291 \text{ mL})}{(2.14 \text{ atm})} = 146 \text{ mL}$

b. $P_1 = 755 \text{ mm Hg}$ $\qquad\qquad\qquad$ $P_2 = 3.51 \text{ atm} = 2668 \text{ mm Hg}$

$\qquad\qquad$ $V_1 = 1.25 \text{ L}$ $\qquad\qquad\qquad\quad$ $V_2 = ? \text{ L}$

$\qquad\qquad$ $V_2 = \dfrac{P_1V_1}{P_2} = \dfrac{(755 \text{ mm Hg})(1.25 \text{ L})}{(2668 \text{ mm Hg})} = 0.354 \text{ L}$

c. $P_1 = 101.4 \text{ kPa} = 760.6 \text{ mm Hg}$ $\qquad$ $P_2 = ? \text{ mm Hg}$

$\qquad\qquad$ $V_1 = 2. 71 \text{ L}$ $\qquad\qquad\qquad\qquad\qquad$ $V_2 = 3.00 \text{ L}$

$\qquad\qquad$ $P_2 = \dfrac{P_1V_1}{V_2} = \dfrac{(760.6 \text{ mm Hg})(2.71 \text{ L})}{(3.00 \text{ L})} = 687 \text{ mm Hg}$

22.   If the pressure exerted on the gas in the balloon is decreased, the volume of the gas in the balloon will increase in inverse proportion to the factor by which the pressure was changed. The factor in this example is (1.01 atm/0.562 atm) = 1.80.

24. $P_2 = \dfrac{P_1 V_1}{V_2} = \dfrac{(1.00 \text{ atm}) (27.2 \text{ mL})}{(1.00 \text{ mL})} = 27.2 \text{ atm}$

26. Charles's law states that the volume of an ideal gas sample varies linearly with the absolute temperature of the gas sample. An experiment such as those indicated in Figures 12.7 and 12.8 can be performed to determine absolute zero. The volume of a sample of gas is measured at several convenient temperatures (e.g., between 0° and 100° C) and the data is plotted. The straight line obtained is then extrapolated to the point where the volume of the gas would become zero. The temperature at which the volume of the gas would be predicted to become zero is then absolute zero.

28. $V = kT;\quad V_1/T_1 = V_2/T_2$

30. $V_1 = 525 \text{ mL}$ $V_2 = ? \text{ mL}$

   $T_1 = 25 \text{ °C} = 298 \text{ K}$ $T_2 = 50 \text{ °C} = 323 \text{ K}$

   $V_2 = \dfrac{V_1 T_2}{T_1} = \dfrac{(525 \text{ mL}) (323 \text{ K})}{(298 \text{ K})} = 569 \text{ mL}$

32. a. $V_1 = 73.5 \text{ mL}$ $V_2 = ? \text{ L}$

   $T_1 = 0 \text{ °C} = 273 \text{ K}$ $T_2 = 25 \text{ °C} = 298 \text{ K}$

   $V_2 = \dfrac{V_1 T_2}{T_1} = \dfrac{(73.5 \text{ mL}) (298 \text{ K})}{(273 \text{ K})} = 80.2 \text{ mL}$

   b. $V_1 = 15.2 \text{ L}$ $V_2 = 10.0 \text{ L}$

   $T_1 = 25 \text{ °C} = 298 \text{ K}$ $T_2 = ? \text{ °C}$

   $T_2 = \dfrac{V_2 T_1}{V_1} = \dfrac{(10.0 \text{ L}) (298 \text{ K})}{(15.2 \text{ L})} = 196 \text{ K} = -77 \text{ °C}$

   c. $V_1 = 1.75 \text{ mL}$ $V_2 = ? \text{ mL}$

   $T_1 = 2.3 \text{ K}$ $T_2 = 0 \text{ °C} = 273 \text{ K}$

   $V_2 = \dfrac{V_1 T_2}{T_1} = \dfrac{(1.75 \text{ mL}) (273 \text{ K})}{(2.3 \text{ K})} = 208 \text{ mL } (2.1 \times 10^2 \text{ mL})$

34. a. $V_1 = 2.01 \times 10^2 \text{ L}$ $V_2 = 5.00 \text{ L}$

   $T_1 = 1150 \text{ °C} = 1423 \text{ K}$ $T_2 = ? \text{ °C}$

   $T_2 = \dfrac{V_2 T_1}{V_1} = \dfrac{(5.00 \text{ L}) (1423 \text{ K})}{(2.01 \times 10^2 \text{ L})} = 35.4 \text{ K} = -238 \text{ °C}$

b. V_1 = 44.2 mL V_2 = ? mL

T_1 = 298 K T_2 = 0 K

$$V_2 = \frac{V_1 T_2}{T_1} = \frac{(44.2 \text{ mL})(0 \text{ K})}{(298 \text{ K})} = 0 \text{ mL (0 K is absolute zero)}$$

c. V_1 = 44.2 mL V_2 = ? mL

T_1 = 298 K T_2 = 0 °C = 273 K

$$V_2 = \frac{V_1 T_2}{T_1} = \frac{(44.2 \text{ mL})(273 \text{ K})}{(298 \text{ K})} = 40.5 \text{ mL}$$

36. Remember that the Celsius temperature scale was defined by humans, and the fact that the temperature is "doubling" from 18°C to 36°C means nothing to the molecules in the sample of gas. As the gas is initially at a temperature of 291 K (18°C), its volume will become half as big at 145.5 K (-128°C) at constant pressure.

38. One method of solution (using Charles's Law in the form $V_1/T_1 = V_2/T_2$) is shown in Example 12.6. A second method might make use of Charles's Law in the form $V = kT$. Using the information that the gas thermometer has a volume of 135 mL at 11°C (284 K), we can solve for the value of the proportionality constant k in the formula, and then use this information to calculate the additional temperatures requested.

$V = kT$

$$k = \frac{V}{T} = \frac{135 \text{ mL}}{284 \text{ K}} = 0.4754 \text{ mL/K}$$

$$T = \frac{V}{k} = \frac{V}{0.475 \text{ mL/K}} = 2.104V$$

For 113 mL, $T = 2.104(113) = 238$ K (−35°C)

For 142 mL, $T = 2.104(142) = 299$ K (26°C)

For 155 mL, $T = 2.104(155) = 326$ K (53°C)

For 127 mL, $T = 2.104(127) = 267$ K (−6°C)

40. $V = an;$ $V_1/n_1 = V_2/n_2$

42. V_1 = 12.0 L V_2 = ? L

n_1 = 2.01 g = 0.502 mol n_2 = 6.52 g = 1.63 mol

$$V_2 = \frac{V_1 n_2}{n_1} = \frac{(12.0 \text{ L})(1.63 \text{ mol})}{(0.502 \text{ mol})} = 39.0 \text{ L}$$

44. Avogadro's law states that the volume of a sample of ideal gas is directly proportional to the amount of gas present in the sample. As the second sample of oxygen has a mass exactly 10 times the mass of the first sample, its volume will be exactly 10 times larger (22.4 L).

46. Real gases most closely approach ideal gas behavior under conditions of relatively high temperatures (0°C or higher) and relatively low pressures (1 atm or lower).

48. For an ideal gas, $PV = nRT$ is true under any conditions. Consider a particular sample of gas (so that n remains constant) at a particular fixed pressure (so that P remains constant also). Suppose that at temperature T_1 the volume of the gas sample is V_1. Then for this set of conditions, the ideal gas equation would be given by

$$PV_1 = nRT_1.$$

If we then change the temperature of the gas sample to a new temperature T_2, the volume of the gas sample changes to a new volume V_2. For this new set of conditions, the ideal gas equation would be given by

$$PV_2 = nRT_2.$$

If we make a ratio of these two expressions for the ideal gas equation for this gas sample, and cancel out terms that are constant for this situation (P, n, and R) we get

$$\frac{PV_1}{PV_2} = \frac{nRT_1}{nRT_2}$$

$$\frac{V_1}{V_2} = \frac{T_1}{T_2}$$

This can be rearranged to the familiar form of Charles's law

$$\frac{V_1}{T_1} = \frac{V_2}{T_2}$$

50. a. $P = 782$ mm Hg $= 1.03$ atm

   $T = 27°C = 300$ K

$$V = \frac{nRT}{P} = \frac{(0.210 \text{ mol})(0.08206 \text{ L atm mol}^{-1} \text{ K}^{-1})(300 \text{ K})}{(1.03 \text{ atm})} = 5.02 \text{ L}$$

   b. $V = 644$ mL $= 0.644$ L

$$P = \frac{nRT}{V} = \frac{(0.0921 \text{ mol})(0.08206 \text{ L atm mol}^{-1} \text{ K}^{-1})(303 \text{ K})}{(0.644 \text{ L})} = 3.56 \text{ atm}$$

   $P = 3.56$ atm $= 2.70 \times 10^3$ mm Hg

   c. $P = 745$ mm $= 0.980$ atm

$$T = \frac{PV}{nR} = \frac{(0.980 \text{ atm})(11.2 \text{ L})}{(0.401 \text{ mol})(0.08206 \text{ L atm mol}^{-1} \text{ K}^{-1})} = 334 \text{ K}$$

52.   a. $T = 25°C = 298 \text{ K}$

$$V = \frac{(0.00831 \text{ mol})(0.08206 \text{ L atm mol}^{-1} \text{ K}^{-1})(298 \text{ K})}{(1.01 \text{ atm})} = 0.201 \text{ L}$$

b. $V = 602 \text{ mL} = 0.602 \text{ L}$

$$P = \frac{(8.01 \times 10^{-3} \text{ mol})(0.08206 \text{ L atm mol}^{-1} \text{ K}^{-1})(310 \text{ K})}{(0.602 \text{ L})} = 0.338 \text{ atm}$$

c. $V = 629 \text{ mL} = 0.629 \text{ L}$

$T = 35°C = 308 \text{ K}$

$$n = \frac{(0.998 \text{ atm})(0.629 \text{ L})}{(0.08206 \text{ L atm mol}^{-1} \text{ K}^{-1})(308 \text{ K})} = 2.48 \times 10^{-2} \text{ mol}$$

54.   Molar mass of He = 4.003 g   25°C = 298 K

$$n = \frac{PV}{RT} = \frac{(120. \text{ atm})(25.0 \text{ L})}{(0.08206 \text{ L atm mol}^{-1} \text{ K}^{-1})(298 \text{ K})} = 122.7 \text{ mol}$$

$$122.7 \text{ mol He} \times \frac{4.003 \text{ g He}}{1 \text{ mol He}} = 491 \text{ g He}$$

56.   The number of moles of *any* ideal gas that can be contained in the tank under the given conditions can first be calculated.

$T = 24°C = 297 \text{ K}$

$$n = \frac{PV}{RT} = \frac{(135 \text{ atm})(200 \text{ L})}{(0.08206 \text{ L atm mol}^{-1} \text{ K}^{-1})(297 \text{ K})} = 1.11 \times 10^3 \text{ mol gas}$$

Molar masses: He, 4.003 g; H_2, 2.016 g

$$\text{for He: } 1.11 \times 10^3 \text{ mol He} \times \frac{4.003 \text{ g}}{1 \text{ mol}} = 4.44 \times 10^3 \text{ g He} = 4.44 \text{ kg He}$$

$$\text{for } H_2\text{: } 1.11 \times 10^3 \text{ mol } H_2 \times \frac{2.016 \text{ g}}{1 \text{ mol}} = 2.24 \times 10^3 \text{ g } H_2 = 2.24 \text{ kg } H_2$$

58.   Molar mass of N_2 = 28.02 g

$$30.1 \text{ g } N_2 \times \frac{1 \text{ mol}}{28.02 \text{ g}} = 1.074 \text{ mol } N_2$$

$$T = \frac{PV}{nR} = \frac{(25.2 \text{ atm})(1.00 \text{ L})}{(1.074 \text{ mol})(0.08206 \text{ L atm mol}^{-1} \text{ K}^{-1})} = 286 \text{ K} = 13°C$$

60.   Molar mass of O_2 = 32.00 g

$56.2 \text{ kg} = 5.62 \times 10^4 \text{ g}$

$$5.62 \times 10^4 \text{ g} \times \frac{1 \text{ mol}}{32.00 \text{ g}} = 1.76 \times 10^3 \text{ mol}$$

$T = 21 \text{ °C} = 294 \text{ K}$

$$P = \frac{nRT}{V} = \frac{(1.76 \times 10^3 \text{ mol})(0.08206 \text{ L atm mol}^{-1} \text{ K}^{-1})(294 \text{ K})}{(125 \text{ L})}$$

$P = 340 \text{ atm}$

62. $P_1 = 0.981 \text{ atm}$ $P_2 = 1.15 \text{ atm}$

$V_1 = 125 \text{ mL}$ $V_2 = ? \text{ mL}$

$T_1 = 100 \text{ °C} = 373 \text{ K}$ $T_2 = 25 \text{ °C} = 298 \text{ K}$

$$V_2 = \frac{P_1 V_1 T_2}{P_2 T_1} = \frac{(0.981 \text{ atm})(125 \text{ mL})(298 \text{ K})}{(1.15 \text{ atm})(373 \text{ K})} = 85.2 \text{ mL}$$

64.   Molar mass of H_2 = 2.016 g   255 mL = 0.255 L   100 °C = 373 K

$$2.51 \text{ g } H_2 \times \frac{1 \text{ mol}}{2.016 \text{ g}} = 1.245 \text{ mol } H_2$$

$$P = \frac{nRT}{V} = \frac{(1.245 \text{ mol})(0.08206 \text{ L atm mol}^{-1} \text{ K}^{-1})(373 \text{ K})}{(0.255 \text{ L})} = 150. \text{ atm}$$

66.   As a gas is bubbled through water, the bubbles of gas become saturated with water vapor, thus forming a gaseous mixture. The total pressure in a sample of gas that has been collected by bubbling through water is made up of two components: the pressure of the gas of interest and the pressure of water vapor. The partial pressure of the gas of interest is then the total pressure of the sample minus the vapor pressure of water.

68.   The volume of a sample of an ideal gas (at a given temperature and pressure) depends on the total number of moles of gas present, not on the specific identity of the gas. For a gaseous mixture as in this problem, the volume of the gas will depend on the total *combined* moles of nitrogen, oxygen, and helium.

Molar masses: N_2, 28.02 g; O_2, 32.00 g; He, 4.003 g

$T = 28 \text{°C} = 301 \text{ K}$

$$6.91 \text{ g } N_2 \times \frac{1 \text{ mol}}{28.02 \text{ g}} = 0.2466 \text{ mol } N_2$$

$$4.71 \text{ g } O_2 \times \frac{1 \text{ mol}}{32.00 \text{ g}} = 0.1472 \text{ mol } O_2$$

$$2.95 \text{ g He} \times \frac{1 \text{ mol}}{4.003 \text{ g}} = 0.7369 \text{ mol He}$$

$n_{\text{total}} = 0.2466 \text{ mol} + 0.1472 \text{ mol} + 0.7369 \text{ mol} = 1.131 \text{ mol}$

$$V = \frac{nRT}{P} = \frac{(1.131 \text{ mol})(0.08206 \text{ L atm mol}^{-1} \text{ K}^{-1})(301 \text{ K})}{(1.05 \text{ atm})} = 26.6 \text{ L}$$

70.   Molar masses: O_2, 32.00 g; N_2, 28.02 g

5.21 kg = 5.21×10^3 g   4.49 kg = 4.49×10^3 g

4.49×10^3 g $O_2 \times \dfrac{1 \text{ mol}}{32.00 \text{ g}}$ = 140. mol O_2

5.21×10^3 g $N_2 \times \dfrac{1 \text{ mol}}{28.02 \text{ g}}$ = 186 mol N_2

n_{total} = 140. mol + 186 mol = 326 mol

T = 24 °C = 297 K

$P = \dfrac{nRT}{V} = \dfrac{(326 \text{ mol})(0.08206 \text{ L atm mol}^{-1} \text{ K}^{-1})(297 \text{ K})}{(50.0 \text{ L})}$ = 159 atm

72.   The pressures must be expressed in the same units, either mm Hg or atm.

$P_{oxygen} = P_{total} - P_{water\ vapor}$

1.02 atm = 775 mm Hg

P_{oxygen} = 775 mm Hg − 23.756 mm Hg = 751.244 mm Hg = 751 mm Hg

74.   1.032 atm = 784.3 mm Hg

$P_{hydrogen}$ = 784.3 mm Hg − 32 mm Hg = 752.3 mm Hg = 0.990 atm

V = 240 mL = 0.240 L

T = 30°C + 273 = 303 K

$n_{hydrogen} = P_{hydrogen}V/RT = \dfrac{(0.990 \text{ atm})(0.240 \text{ L})}{(0.08206 \text{ L atm mol}^{-1} \text{ K}^{-1})(303 \text{ K})}$ = 0.00956 mol

0.00956 mol $H_2 \times \dfrac{1 \text{ mol Zn}}{1 \text{ mol H}_2}$ = 0.00956 mol of Zn must have reacted

molar mass of Zn = 65.38 g

0.00956 mol Zn $\times \dfrac{65.38 \text{ g Zn}}{1 \text{ mol Zn}}$ = 0.625 g Zn must have reacted

76.   A theory is successful if it explains known experimental observations. Theories that have been successful in the past may not be successful in the future (for example, as technology evolves, more sophisticated experiments may be possible in the future).

78.   Chemists believe the pressure exerted by a gas sample on the walls of its container arises from collisions between the gas molecules and the walls of the container.

80.   no

82.   If the temperature of a sample of gas is increased, the average kinetic energy of the particles of gas increases. This means that the speeds of the particles increase. If the particles have a higher speed, they will hit the walls of the container more frequently and with greater force, thereby increasing the pressure.

84.   Standard Temperature and Pressure, STP = 0°C, 1 atm pressure. These conditions were chosen because they are easy to attain and reproduce *experimentally*. The barometric pressure within a laboratory is likely to be near 1 atm most days, and 0°C can be attained with a simple ice bath.

86.   Molar mass of Al = 26.98 g

$$1.55 \text{ g Al} \times \frac{1 \text{ mol}}{26.98 \text{ g}} = 0.0574 \text{ mol Al}$$

From the balanced chemical equation, the amount of oxygen needed is

$$0.0574 \text{ mol Al} \times \frac{3 \text{ mol O}_2}{2 \text{ mol Al}} = 0.0861 \text{ mol O}_2$$

$$0.0861 \text{ mol O}_2 \times \frac{22.4 \text{ L}}{1 \text{ mol}} = 1.93 \text{ L.}$$

88.   Molar mass of Mg = 24.31 g

$$1.02 \text{ g Mg} \times \frac{1 \text{ mol}}{24.31 \text{ g}} = 0.0420 \text{ mol Mg}$$

As the coefficients for Mg and Cl_2 in the balanced equation are the same, for 0.0420 mol of Mg reacting we will need 0.0420 mol of Cl_2.

STP: 1.00 atm, 273 K

$$V = \frac{nRT}{P} = \frac{(0.0420 \text{ mol})(0.08206 \text{ L atm mol}^{-1} \text{ K}^{-1})(273 \text{ K})}{(1.00 \text{ atm})} = 0.941 \text{ L Cl}_2$$

90.   Molar mass of $CaCO_3$ = 100.08 g

$$4.74 \text{ g CaCO}_3 \times \frac{1 \text{ mol}}{100.08 \text{ g}} = 0.0474 \text{ mol CaCO}_3$$

As the coefficients of $CaCO_3$ and CO_2 in the balanced chemical equation are the same, when 0.0474 mol $CaCO_3$ is heated, 0.0474 mol CO_2 results.

T = 26 °C = 299 K

$$V = \frac{nRT}{P} = \frac{(0.0474 \text{ mol})(0.08206 \text{ L atm mol}^{-1} \text{ K}^{-1})(299 \text{ K})}{(0.997 \text{ atm})} = 1.17 \text{ L CO}_2$$

92. Molar mass of Mg_3N_2 = 100.95 g

$$10.3 \text{ g } Mg_3N_2 \times \frac{1 \text{ mol}}{100.95 \text{ g}} = 0.102 \text{ mol } Mg_3N_2$$

From the balanced chemical equation, the amount of NH_3 produced will be

$$0.102 \text{ mol } Mg_3N_2 \times \frac{2 \text{ mol } NH_3}{1 \text{ mol } Mg_3N_2} = 0.204 \text{ mol } NH_3$$

$T = 24°C = 297$ K $\qquad\qquad$ $P = 752$ mm Hg $= 0.989$ atm

$$V = \frac{nRT}{P} = \frac{(0.204 \text{ mol})(0.08206 \text{ L atm mol}^{-1} \text{ K}^{-1})(297 \text{ K})}{(0.989 \text{ atm})} = 5.03 \text{ L}$$

This assumes that the ammonia was collected dry.

94. Molar masses: O_2, 32.00 g; N_2, 28.02 g

$$26.2 \text{ g } O_2 \times \frac{1 \text{ mol}}{32.00 \text{ g}} = 0.819 \text{ mol } O_2$$

$$35.1 \text{ g } N_2 \times \frac{1 \text{ mol}}{28.02 \text{ g}} = 1.25 \text{ mol } N_2$$

$n_{total} = 0.819 \text{ mol} + 1.25 \text{ mol} = 2.07 \text{ mol}$

$T = 35°C = 308$ K

$P = 755$ mm Hg $= 0.993$ atm

$$V = \frac{nRT}{P} = \frac{(2.07 \text{ mol})(0.08206 \text{ L atm mol}^{-1} \text{ K}^{-1})(308 \text{ K})}{(0.993 \text{ atm})} = 52.7 \text{ L}$$

96. $P_1 = 1.47$ atm $\qquad\qquad$ $P_2 = 1.00$ atm (Standard Pressure)

$V_1 = 145$ mL $\qquad\qquad$ $V_2 = ?$ mL

$T_1 = 44°C = 317$ K $\qquad\qquad$ $T_2 = 0°C = 273$ K (Standard Temperature)

$$V_2 = \frac{P_1 V_1 T_2}{P_2 T_1} = \frac{(1.47 \text{ atm})(145 \text{ mL})(273 \text{ K})}{(1.00 \text{ atm})(317 \text{ K})} = 184 \text{ mL}$$

98. Molar masses: He, 4.003 g; Ne, 20.18 g

$$6.25 \text{ g He} \times \frac{1 \text{ mol}}{4.003 \text{ g}} = 1.561 \text{ mol He}$$

$$4.97 \text{ g Ne} \times \frac{1 \text{ mol}}{20.18 \text{ g}} = 0.2463 \text{ mol Ne}$$

$n_{total} = 1.561 \text{ mol} + 0.2463 \text{ mol} = 1.807 \text{ mol}$

As 1 mol of an ideal gas occupies 22.4 L at STP, the volume is given by

$$1.807 \text{ mol} \times \frac{22.4 \text{ L}}{1 \text{ mol}} = 40.48 \text{ L} = 40.5 \text{ L}.$$

The partial pressure of a given gas in a mixture will be proportional to what *fraction* of the total number of moles of gas the given gas represents

$$P_{He} = \frac{1.561 \text{ mol He}}{1.807 \text{ mol total}} \times 1.00 \text{ atm} = 0.8639 \text{ atm} = 0.864 \text{ atm}$$

$$P_{Ne} = \frac{0.2463 \text{ mol Ne}}{1.807 \text{ mol total}} \times 1.00 \text{ atm} = 0.1363 \text{ atm} = 0.136 \text{ atm}$$

100. Molar mass of Ca = 40.08 g

$$4.15 \text{ g Ca} \times \frac{1 \text{ mol}}{40.08 \text{ g}} = 0.1035 \text{ mol Ca}$$

$$0.1034 \text{ mol Ca} \times \frac{1 \text{ mol Cl}_2}{1 \text{ mol Ca}} = 0.1034 \text{ mol Cl}_2$$

$$V = nRT/P = \frac{(0.1034)(0.08206 \text{ L atm mol}^{-1} \text{ K}^{-1})(298 \text{ K})}{(1.02 \text{ atm})} = 2.48 \text{ L}$$

102. 125 mL = 0.125 L

$$0.125 \text{ L} \times \frac{1 \text{ mol}}{22.4 \text{ L}} = 0.00558 \text{ mol H}_2$$

From the balanced chemical equation, one mole of zinc is required for each mole of hydrogen produced. Therefore, 0.00558 mol of Zn will be required.

$$0.00558 \text{ mol Zn} \times \frac{65.38 \text{ g Zn}}{1 \text{ mol Zn}} = 0.365 \text{ g Zn}$$

104. twice

106. a. $PV = k; \ P_1V_1 = P_2V_2$

b. $V = kT; \ V_1/T_1 = V_2/T_2$

c. $V = an; \ V_1/n_1 = V_2/n_2$

d. $PV = nRT$

e. $P_1V_1/T_1 = P_2V_2/T_2$

108. First determine what volume the helium in the tank would have if it were at a pressure of 755 mm Hg (corresponding to the pressure the gas will have in the balloons).

8.40 atm = 6384 mm Hg

$$V_2 = (25.2 \text{ L}) \times \frac{6384 \text{ mm Hg}}{755 \text{ mm Hg}} = 213 \text{ L}$$

Allowing for the fact that 25.2 L of He will have to remain in the tank, this leaves 213 - 25.2 = 187.8 L of He for filling the balloons.

$$187.8 \text{ L He} \times \frac{1 \text{ balloon}}{1.50 \text{ L He}} = 125 \text{ balloons}$$

110. According to the balanced chemical equation, when 1 mol of $(NH_4)_2CO_3$ reacts, a total of 4 moles of gaseous substances are produced.

molar mass $(NH_4)_2CO_3$ = 96.09 g

$$52.0 \text{ g} \times \frac{1 \text{ mol}}{96.09 \text{ g}} = 0.541 \text{ mol}$$

As 0.541 mol of $(NH_4)_2CO_3$ reacts, 4(0.541) = 2.16 mol of gaseous products result.

453 °C = 726 K

$$V = \frac{(2.16 \text{ mol})(0.08206 \text{ L atm/mol}^{-1} \text{ K}^{-1})(726 \text{ K})}{(1.04 \text{ atm})} = 124 \text{ L}$$

112. $CaCO_3(s) + 2H^+(aq) \rightarrow Ca^{2+}(aq) + H_2O(l) + CO_2(g)$

molar mass $CaCO_3$ = 100.1 g

$$10.0 \text{ g CaCO}_3 \times \frac{1 \text{ mol CaCO}_3}{100.1 \text{ g CaCO}_3} = 0.0999 \text{ mol CaCO}_3 = 0.0999 \text{ mol CO}_2 \text{ also}$$

60°C + 273 = 333 K

$P_{\text{carbon dioxide}} = P_{\text{total}} - P_{\text{water vapor}}$

$P_{\text{carbon dioxide}}$ = 774 mm Hg - 149.4 mm Hg = 624.6 mm Hg = 0.822 atm

$$V_{\text{wet}} = \frac{(0.0999 \text{ mol})(0.08206 \text{ L atm mol}^{-1} \text{ K}^{-1})(333 \text{ K})}{(0.822 \text{ atm})} = 3.32 \text{ L wet CO}_2$$

$$V_{\text{dry}} = 3.32 \text{ L} \times \frac{624.6 \text{ mm Hg}}{774 \text{ mm Hg}} = 2.68 \text{ L}$$

114. $2KClO_3(s) \rightarrow 2KCl(s) + 3O_2(g)$

molar mass $KClO_3$ = 122.6 g

$$50.0 \text{ g KClO}_3 \times \frac{1 \text{ mol KClO}_3}{122.6 \text{ g KClO}_3} = 0.408 \text{ mol KClO}_3$$

$$0.408 \text{ mol KClO}_3 \times \frac{3 \text{ mol O}_2}{2 \text{ mol KClO}_3} = 0.612 \text{ mol O}_2$$

25°C + 273 = 298 K   630. torr = 0.829 atm

$$V = nRT/P = \frac{(0.612 \text{ mol})(0.08206 \text{ L atm mol}^{-1} \text{ K}^{-1})(298 \text{ K})}{(0.829 \text{ atm})} = 18.1 \text{ L O}_2$$

116. a. $752 \text{ mm Hg} \times \dfrac{101{,}325 \text{ Pa}}{760 \text{ mm Hg}} = 1.00 \times 10^5 \text{ Pa}$

  b. $458 \text{ kPa} \times \dfrac{1 \text{ atm}}{101.325 \text{ kPa}} = 4.52 \text{ atm}$

  c. $1.43 \text{ atm} \times \dfrac{760 \text{ mm Hg}}{1 \text{ atm}} = 1.09 \times 10^3 \text{ mm Hg}$

  d. $842 \text{ torr} = 842 \text{ mm Hg}$

118. a. $645 \text{ mm Hg} \times \dfrac{1 \text{ atm}}{760 \text{ mm Hg}} \times \dfrac{101{,}325 \text{ Pa}}{1 \text{ atm}} = 8.60 \times 10^4 \text{ Pa}$

  b. $221 \text{ kPa} = 221 \times 10^3 \text{ Pa} = 2.21 \times 10^5 \text{ Pa}$

  c. $0.876 \text{ atm} \times \dfrac{101{,}325 \text{ Pa}}{1 \text{ atm}} = 8.88 \times 10^4 \text{ Pa}$

  d. $32 \text{ torr} \times \dfrac{1 \text{ atm}}{760 \text{ torr}} \times \dfrac{101{,}325 \text{ Pa}}{1 \text{ atm}} = 4.3 \times 10^3 \text{ Pa}$

120. a. $1.00 \text{ mm Hg} = 1.00 \text{ torr}$

  $V = 255 \text{ mL} \times \dfrac{1.00 \text{ torr}}{2.00 \text{ torr}} = 128 \text{ mL}$

  b. $1.0 \text{ atm} = 101.325 \text{ kPa}$

  $V = 1.3 \text{ L} \times \dfrac{1.0 \text{ kPa}}{101.325 \text{ kPa}} = 1.3 \times 10^{-2} \text{ L}$

  c. $1.0 \text{ mm Hg} = 0.133 \text{ kPa}$

  $V = 1.3 \text{ L} \times \dfrac{1.0 \text{ kPa}}{0.133 \text{ kPa}} = 9.8 \text{ L}$

122. $1.52 \text{ L} = 1.52 \times 10^3 \text{ mL}$

  $755 \text{ mm Hg} \times \dfrac{1.52 \times 10^3 \text{ mL}}{450 \text{ mL}} = 2.55 \times 10^3 \text{ mm Hg}$

124. a. $74°C + 273 = 347 \text{ K}$ $-74°C + 273 = 199 \text{ K}$

  $100. \text{ mL} \times \dfrac{199 \text{ K}}{347 \text{ K}} = 57.3 \text{ mL}$

  b. $100°C + 273 = 373 \text{ K}$

  $373 \text{ K} \times \dfrac{600 \text{ mL}}{500 \text{ mL}} = 448 \text{ K} \ (175°C)$

c.   zero (the volume of any gas sample becomes zero at 0 K)

126.   12 °C + 273 = 285 K   192 °C + 273 = 465 K

$$75.2 \text{ mL} \times \frac{465 \text{ K}}{285 \text{ K}} = 123 \text{ mL}$$

128.   For a given gas, the number of moles present in a sample is directly proportional to the mass of the sample. The problem therefore can be solved even though the gas is not identified (so that its molar mass is not known).

$$23.2 \text{ g} \times \frac{10.4 \text{ L}}{93.2 \text{ L}} = 2.59 \text{ g}$$

130.   a. $V = 21.2 \text{ mL} = 0.0212 \text{ L}$

$$T = PV/nR = \frac{(1.034 \text{ atm})(0.0212 \text{ L})}{(0.00432 \text{ mol})(0.08206 \text{ L atm mol}^{-1} \text{ K}^{-1})} = 61.8 \text{ K}$$

b. $V = 1.73 \text{ mL} = 0.00173 \text{ L}$

$$P = nRT/V = \frac{(0.000115 \text{ mol})(0.08206 \text{ L atm mol}^{-1} \text{ K}^{-1})(182 \text{ K})}{(0.00173 \text{ L})}$$

$P = 0.993 \text{ atm}$

c. $P = 1.23 \text{ mm Hg} = 0.00162 \text{ atm}$

$T = 152°\text{C} + 273 = 425 \text{ K}$

$$V = nRT/P = \frac{(0.773 \text{ mol})(0.08206 \text{ L atm mol}^{-1} \text{ K}^{-1})(425 \text{ K})}{(0.00162 \text{ atm})}$$

$V = 1.66 \times 10^4 \text{ L}$

132. $27°\text{C} + 273 = 300 \text{ K}$

The number of moles of gas it takes to fill the 100. L tanks to 120 atm at 27°C is independent of the identity of the gas.

$$n = PV/RT = \frac{(120 \text{ atm})(100. \text{ L})}{(0.08206 \text{ L atm mol}^{-1} \text{ K}^{-1})(300 \text{ K})} = 487 \text{ mol}$$

487 mol of *any* gas will fill the tanks to the required specifications.

molar masses: CH_4, 16.0 g; N_2, 28.0 g; CO_2, 44.0 g

for CH_4: (487 mol)(16.0 g/mol) = 7792 g = 7.79 kg CH_4

for N_2:   (487 mol)(28.0 g/mol) = 13,636 g = 13.6 kg N_2

for CO_2: (487 mol)(44.0 g/mol) = 21,428 g = 21.4 kg CO_2

134.   molar mass of O_2 = 32.00 g

55 mg = 0.055 g

$n = 0.055 \text{ g} \times \dfrac{1 \text{ mol}}{32.00 \text{ g}} = 0.0017 \text{ mol}$

V = 100. mL = 0.100 L

T = 26°C + 273 = 299 K

$P = nRT/V = \dfrac{(0.0017 \text{ mol})(0.08206 \text{ L atm mol}^{-1} \text{ K}^{-1})(299 \text{ K})}{(0.100 \text{ L})} = 0.42 \text{ atm}$

136. P_1 = 1.13 atm P_2 = 1.89 atm

V_1 = 100 mL = 0.100 L V_2 = 500 mL = 0.500 L

T_1 = 300 K T_2 = ?

$T_2 = \dfrac{T_1 P_2 V_2}{P_1 V_1} = \dfrac{(300 \text{ K})(1.89 \text{ atm})(0.500 \text{ L})}{(1.13 \text{ atm})(0.100 \text{ L})} = 2.51 \times 10^3 \text{ K}$

Note that the calculation could have been carried through with the two volumes expressed in milliliters because the universal gas constant does not appear explicitly in this form of the ideal gas equation.

138.   molar masses: N_2, 28.0 g; He, 4.003 g

$12.1 \text{ g N}_2 \times \dfrac{1 \text{ mol N}_2}{28.0 \text{ g N}_2} = 0.432 \text{ mol N}_2$

$4.05 \text{ g He} \times \dfrac{1 \text{ mol He}}{4.003 \text{ g He}} = 1.01 \text{ mol He}$

Total moles of gas = 0.432 mol + 1.01 mol = 1.44 mol

STP: 1.00 atm, 273 K

$V = nRT/P = \dfrac{(1.44 \text{ mol})(0.08206 \text{ L atm mol}^{-1} \text{ K}^{-1})(273 \text{ K})}{(1.00 \text{ atm})} = 32.3 \text{ L}$

140.   molar mass of NH_3 = 17.03 g

$5.00 \text{ g NH}_3 \times \dfrac{1 \text{ mol NH}_3}{17.03 \text{ g NH}_3} = 0.294 \text{ mol NH}_3 \text{ to be produced}$

$N_2(g) + 3H_2(g) \rightarrow 2NH_3(g)$

$0.294 \text{ mol NH}_3 \times \dfrac{1 \text{ mol N}_2}{2 \text{ mol NH}_3} = 0.147 \text{ mol N}_2 \text{ required}$

$$0.294 \text{ mol NH}_3 \times \frac{3 \text{ mol H}_2}{2 \text{ mol NH}_3} = 0.441 \text{ mol H}_2 \text{ required}$$

$$11°C + 273 = 284 \text{ K}$$

$$V_{\text{nitrogen}} = \frac{(0.147 \text{ mol})(0.08206 \text{ L atm mol}^{-1} \text{ K}^{-1})(284 \text{ K})}{(0.998 \text{ atm})} = 3.43 \text{ L N}_2$$

$$V_{\text{hydrogen}} = \frac{(0.441 \text{ mol})(0.08206 \text{ L atm mol}^{-1} \text{ K}^{-1})(284 \text{ K})}{(0.998 \text{ atm})} = 10.3 \text{ L H}_2$$

142. $2Cu_2S(s) + 3O_2(g) \rightarrow 2Cu_2O(s) + 2SO_2(g)$

molar mass Cu_2S = 159.2 g

$$25 \text{ g Cu}_2\text{S} \times \frac{1 \text{ mol Cu}_2\text{S}}{159.2 \text{ g Cu}_2\text{S}} = 0.1570 \text{ mol Cu}_2\text{S}$$

$$0.1570 \text{ mol Cu}_2\text{S} \times \frac{3 \text{ mol O}_2}{2 \text{ mol Cu}_2\text{S}} = 0.2355 \text{ mol O}_2$$

$$27.5°C + 273 = 301 \text{ K}$$

$$V_{\text{oxygen}} = \frac{(0.2355 \text{ mol})(0.08206 \text{ L atm mol}^{-1} \text{ K}^{-1})(301 \text{ K})}{(0.998 \text{ atm})} = 5.8 \text{ L O}_2$$

$$0.1570 \text{ mol Cu}_2\text{S} \times \frac{2 \text{ mol SO}_2}{2 \text{ mol Cu}_2\text{S}} = 0.1570 \text{ mol SO}_2$$

$$V_{\text{sulfur dioxide}} = \frac{(0.1570 \text{ mol})(0.08206 \text{ L atm mol}^{-1} \text{ K}^{-1})(301 \text{ K})}{(0.998 \text{ atm})}$$

$$V_{\text{sulfur dioxide}} = 3.9 \text{ L SO}_2$$

144. One mole of any ideal gas occupies 22.4 L at STP.

$$35 \text{ mol N}_2 \times \frac{22.4 \text{ L}}{1 \text{ mol}} = 7.8 \times 10^2 \text{ L}$$

146. molar masses: He, 4.003 g; Ar, 39.95 g; Ne, 20.18 g

$$5.0 \text{ g He} \times \frac{1 \text{ mol He}}{4.003 \text{ g He}} = 1.249 \text{ mol He}$$

$$1.0 \text{ g Ar} \times \frac{1 \text{ mol Ar}}{39.95 \text{ g Ar}} = 0.02503 \text{ mol Ar}$$

$$3.5 \text{ g Ne} \times \frac{1 \text{ mol Ne}}{20.18 \text{ g Ne}} = 0.1734 \text{ mol Ne}$$

Total moles of gas = 1.249 + 0.02503 + 0.1734 = 1.447 mol

22.4 L is the volume occupied by one mole of any ideal gas at STP. This would apply even if the gas sample is a *mixture* of individual gases.

$$1.447 \text{ mol} \times \frac{22.4 \text{ L}}{1 \text{ mol}} = 32 \text{ L total volume for the mixture}$$

The *partial pressure* of each individual gas in the mixture will be related to what *fraction* on a mole basis each gas represents in the mixture.

$$P_{He} = 1.00 \text{ atm} \times \frac{1.249 \text{ mol He}}{1.447 \text{ mol total}} = 0.86 \text{ atm}$$

$$P_{Ar} = 1.00 \text{ atm} \times \frac{0.02503 \text{ mol Ar}}{1.447 \text{ mol total}} = 0.017 \text{ atm}$$

$$P_{Ne} = 1.00 \text{ atm} \times \frac{0.1734 \text{ mol Ne}}{1.447 \text{ mol total}} = 0.12 \text{ atm}$$

148.   The solution is only 50% H_2O_2. Therefore 125 g solution = 62.5 g H_2O_2

molar mass of H_2O_2 = 34.02 g

$$62.5 \text{ g } H_2O_2 \times \frac{1 \text{ mol}}{34.02 \text{ g}} = 1.84 \text{ mol } H_2O_2$$

$$1.84 \text{ mol } H_2O_2 \times \frac{1 \text{ mol } O_2}{2 \text{ mol } H_2O_2} = 0.920 \text{ mol } O_2$$

$T = 27 \text{ °C} = 300 \text{ K}$ $\qquad\qquad$ $P = 764 \text{ mm Hg} = 1.01 \text{ atm}$

$$V = nRT/P = \frac{(0.920)(0.08206 \text{ L atm mol}^{-1} \text{ K}^{-1})(300 \text{ K})}{(1.01 \text{ atm})} = 22.4 \text{ L}$$

Chapter 13   Liquids and Solids

2.   Water exerts its cooling effect in nature in many ways. Water, as perspiration, helps cool the human body (the evaporation of water from skin is an endothermic process; the heat required for evaporation comes from the body). Large bodies of natural water (e.g., the oceans) have a cooling effect on nearby land masses (the interior of the United States, away from the oceans, tends to be hotter than coastal regions). In industry, water is used as a coolant in *many* situations. Some nuclear power plants, for example, use water to cool the reactor core. Many office buildings are air-conditioned in summer by circulating cold water systems.

4.   The fact that water expands when it freezes often results in broken water pipes during cold weather. The expansion of water when it freezes also makes ice float on liquid water. The expansion of a given mass of water into a larger volume upon freezing lowers the density of ice compared to liquid water. Aquatic life (and probably all life) could not exist if ice sank in water.

6.   Sloped portions of a heating/cooling curve represent *changes in temperature* as heat is applied or removed; for example, in the cooling/heating curve shown, there are sloped portions representing the heating of ice, the heating of liquid water, and the heating of steam, as heat continues to be applied. Flat portions of such curves represent equilibrium transitions between states; for example, the flat portions in the curve shown represent the solid-liquid (melting-freezing) transition and the liquid-vapor (boiling-condensation) phase transitions.

8.   As a liquid is heated, the motions of the molecules increase as the temperature rises. As the liquid reaches its boiling point, bubbles of vapor begin to form in the liquid, which rise to the surface of the liquid and burst. As the liquid remains at its boiling point, the additional heat energy being supplied to the liquid is used to overcome attractive forces among the molecules in the liquid. As heat energy continues to be applied, more and more molecules will be moving in the right direction and with sufficient energy to escape from the liquid.

10.   To melt a solid, or to vaporize a liquid, the molecules of the substance must be moved apart: thus, it is the *inter*molecular forces that must be overcome.

12.   Steam at 100 °C contains more energy than liquid water at 100°C because the molecules have absorbed an additional amount of energy (the heat of vaporization) in forming the gaseous state.

14. $10.0 \text{ g X} \times \dfrac{1 \text{ mol X}}{52 \text{ g X}} = 0.192 \text{ mol X}$

$0.192 \text{ mol} \times \dfrac{2.5 \text{ kJ}}{1 \text{ mol}} = 0.48 \text{ kJ}$

$$25.0 \text{ g X} \times \frac{1 \text{ mol X}}{52 \text{ g X}} = 0.481 \text{ mol X}$$

$$0.481 \text{ mol} \times \frac{55.3 \text{ kJ}}{1 \text{ mol}} = 27 \text{ kJ}$$

16. molar mass H_2O = 18.02 g

$$25.0 \text{ g } H_2O \times \frac{1 \text{ mol } H_2O}{18.02 \text{ g } H_2O} = 1.39 \text{ mol}$$

To melt the ice: $1.39 \text{ mol} \times \dfrac{6.02 \text{ kJ}}{1 \text{ mol}} = 8.37 \text{ kJ}$

$$37.5 \text{ g } H_2O \times \frac{1 \text{ mol } H_2O}{18.02 \text{ g } H_2O} = 2.08 \text{ mol}$$

To vaporize the liquid: $2.08 \text{ mol} \times \dfrac{40.6 \text{ kJ}}{1 \text{ mol}} = 84.4 \text{ kJ}$

To heat the liquid: $55.2 \text{ g} \times 4.18 \dfrac{J}{g \text{ °C}} \times 100\text{°C} = 23{,}073 \text{ J} = 23.1 \text{ kJ}$

18. The *molar* heat of fusion of aluminum is the heat required to melt 1 mol.

$$\frac{3.95 \text{ kJ}}{1 \text{ g}} \times \frac{26.98 \text{ g}}{1 \text{ mol}} = 107 \text{ kJ/mol}$$

$$10.0 \text{ g} \times \frac{3.95 \text{ kJ}}{1 \text{ g}} = 39.5 \text{ kJ required to melt 10.0 g Al}$$

$$10.0 \text{ mol} \times \frac{107 \text{ kJ}}{1 \text{ mol}} = 1.07 \times 10^3 \text{ kJ required to melt 10.0 mol Al.}$$

20. Dipole-dipole forces are relatively stronger at short distances; they are short-range forces. Molecules must first closely approach one another before dipole-dipole forces can cause attraction between molecules.

22. Water molecules are able to form strong *hydrogen bonds* with each other. These bonds are an especially strong form of dipole-dipole forces and are only possible when hydrogen atoms are bonded to the most electronegative elements (N, O, and F). The particularly strong intermolecular forces in H_2O require much higher temperatures (high energies) to be overcome in order to permit the liquid to boil. We take the fact that water has a much higher boiling point than the other hydrogen compounds of the Group 6 elements as proof that a special force is at play in water (hydrogen bonding).

24.   The fact that such nonpolar, monatomic atoms *can* be liquefied and solidified indicates that there must be *some* sort of intermolecular forces possible between atoms in these substances. London dispersion forces arise when a temporary (instantaneous) dipolar arrangement of charge develops as the electrons of an atom move around its nucleus. This instantaneous dipole can induce a similar dipole in a neighboring atom, leading to a momentary attraction.

26.   a.   London dispersion forces (nonpolar, atoms)

      b.   London dispersion forces (nonpolar molecules)

      c.   dipole-dipole forces; London dispersion forces

      d.   hydrogen bonding (H attached to O); London dispersion forces

28.   An increase in the heat of fusion is observed for an increase in the size of the halogen atom involved (the electron cloud of a larger atom is more easily polarized by an approaching dipole, thus giving larger London dispersion forces).

30.   For a homogeneous mixture to be able to form at all, the forces between molecules of the two substances being mixed must be at least *comparable in magnitude* to the intermolecular forces within each *separate* substance. Apparently in the case of a water-ethanol mixture, the forces that exist when water and ethanol are mixed are stronger than water-water or ethanol-ethanol forces in the separate substances. This allows ethanol and water molecules to approach each other more closely in the mixture than either substance's molecules could approach a like molecule in separate substances. There is strong hydrogen bonding in both ethanol and water.

32.   Vapor pressure is the pressure of vapor present *at equilibrium* above a liquid in a sealed container at a particular temperature. When a liquid is placed in a closed container, molecules of the liquid evaporate freely into the empty space above the liquid. As the number of molecules present in the vapor state increases with time, vapor molecules begin to rejoin the liquid state (condense). Eventually a dynamic equilibrium is reached between evaporation and condensation in which the net number of molecules present in the vapor phase becomes *constant* with time.

34.   A method is shown in Figure 13.10. The apparatus consists of a barometer into which a volatile liquid may be injected. As mercury is more dense than other liquids, the injected volatile liquid rises to the top of the mercury column in the tube and floats on top of the mercury. The space above the mercury is a vacuum, the volatile liquid evaporates into the empty space. As the liquid is converted to the gaseous state, the level of the mercury column drops as the pressure of vapor builds.

36.   a.   HF   Although both substances are capable of hydrogen bonding, water has two O-H bonds that can be involved in hydrogen bonding versus only one F-H bond in HF.

b. CH_3OCH_3 As there is no H attached to the O atom, no hydrogen bonding can exist. Moreover, as there is no hydrogen bonding possible, the molecule should be relatively more volatile than CH_3CH_2OH even though it contains the same number of atoms of each element.

c. CH_3SH Hydrogen bonding is not as important for a S–H bond (because S has a lower electronegativity than O). As there is little hydrogen bonding, the molecule is relatively more volatile than CH_3OH.

38.   Both substances have the same molar mass. However, ethyl alcohol contains a hydrogen atom directly bonded to an oxygen atom. Therefore, hydrogen bonding can exist in ethyl alcohol, whereas only weak dipole-dipole forces can exist in dimethyl ether. Dimethyl ether is more volatile; ethyl alcohol has a higher boiling point.

40.   *Ionic* solids have as their fundamental particles positive and negative *ions*; a simple example is sodium chloride, in which Na^+ and Cl^- ions are held together by strong electrostatic forces.

Molecular solids have molecules as their fundamental particles, with the molecules being held together in the crystal by dipole-dipole forces, hydrogen bonding forces, or London dispersion forces (depending on the identity of the substance); simple examples of molecular solids include ice (H_2O) and ordinary table sugar (sucrose).

Atomic solids have simple atoms as their fundamental particles, with the atoms being held together in the crystal either by covalent bonding (as in graphite or diamond) or by metallic bonding (as in copper or other metals).

42.   The interparticle forces in ionic solids (the ionic bond) are much stronger than the interparticle forces in molecular solids (dipole-dipole forces, London forces, etc.). The difference in intermolecular forces is most clearly shown in the great differences in melting points and boiling points between ionic and molecular solids. For example, table salt and ordinary sugar are both crystalline solids that appear very similar. Yet sugar can be melted easily in a saucepan during the making of candy, whereas even the full heat of a stove will not melt salt.

44.   Ionic solids consist of a crystal lattice of basically alternating positively and negatively charged ions. A given ion is surrounded by several ions of the opposite charge, all of which electrostatically attract it strongly. This pattern repeats itself throughout the crystal. The existence of these strong electrostatic forces throughout the crystal means a great deal of energy must be applied to overcome the forces and melt the solid.

46. Ordinary ice contains nonlinear, highly polar water molecules. In addition, there is extensive, strong hydrogen bonding possible between water molecules in ordinary ice. Dry ice, on the other hand, consists of linear, nonpolar carbon dioxide molecules, and only very weak intermolecular forces are possible.

48. Although ions exist in the solid, liquid, or dissolved states, in the solid state the ions are rigidly held in place in the crystal lattice and cannot *move* so as to conduct an electrical current.

50. The presence of atoms of a second element in a given metal's crystal lattice changes the properties of the metal: frequently the alloy is stronger than either of the original components because the irregularities introduced into the crystal lattice by the presence of a second element's atoms prevent the crystal from being deformed as easily. The properties of iron may be modified by alloying with many different substances, particularly with carbon, nickel, and cobalt. Steels with relatively high carbon content are exceptionally strong, whereas steels with low carbon content are softer, more malleable, and more ductile.

52. j

54. f

56. d

58. a

60. l

62. Dimethyl ether has the larger vapor pressure. No hydrogen bonding is possible because the O atom does not have a hydrogen atom attached. Hydrogen bonding can occur *only* when a hydrogen atom is *directly* attached to a strongly electronegative atom (such as N, O, or F). Hydrogen bonding *is* possible in ethanol (ethanol contains an –OH group).

64. a. H_2. London dispersion forces are the only intermolecular forces present in these nonpolar molecules; typically London forces become larger with increasing atomic size (as the atoms become bigger, the edge of the electron cloud lies farther from the nucleus and becomes more easily distorted).

     b. Xe. Only the relatively weak London forces could exist in a crystal of Xe atoms, whereas in NaCl strong ionic forces exist. In diamond, strong covalent bonding exists between carbon atoms.

     c. Cl_2. Only London forces exist among such nonpolar molecules. London forces become larger with increasing atomic size.

66.   Steel is a general term applied to alloys consisting primarily of iron, but with small amounts of other substances added. Whereas pure iron itself is relatively soft, malleable, and ductile, steels are typically much stronger and harder, and much less subject to damage.

68.   Water is the solvent in which cellular processes take place in living creatures. Water in the oceans moderates the Earth's temperature. Water is used in industry as a cooling agent. Water serves as a means of transportation on the Earth's oceans. The liquid range is 0°C to 100°C at 1 atm pressure.

70.   At higher altitudes, the boiling points of liquids, such as water, are lower because there is a lower atmospheric pressure above the liquid. The temperature at which food cooks is determined by the temperature to which the water in the food can be heated before it escapes as steam. Thus, food cooks at a lower temperature at high elevations where the boiling point of water is lowered.

72.   Heat of fusion (melt); heat of vaporization (boil).
       The heat of vaporization is always larger, because virtually all of the intermolecular forces must be overcome to form a gas. In a liquid, considerable intermolecular forces remain. Thus going from a solid to liquid requires less energy than going from the liquid to the gas.

74.   Dipole-dipole interactions are typically about 1% as strong as a covalent bond. Dipole-dipole interactions represent electrostatic attractions between portions of molecules that carry only a *partial* positive or negative charge. Such forces require the molecules that are interacting to come *near* enough to each other.

76.   London dispersion forces are relatively weak forces that arise among noble gas atoms and in nonpolar molecules. London forces are due to *instantaneous dipoles* that develop when one atom (or molecule) momentarily distorts the electron cloud of another atom (or molecule). London forces are typically weaker than either permanent dipole-dipole forces or covalent bonds.

78.   For every mole of liquid water that evaporates, several kilojoules of heat must be absorbed to provide kinetic energy to overcome attractive forces among the molecules. This heat is absorbed by the water from its surroundings.

80.   In NH_3, strong hydrogen bonding can exist. In CH_4, because the molecule is nonpolar, only the relatively weak London dispersion forces exist.

82.   In a crystal of ice, strong *hydrogen bonding* forces are present, whereas in the crystal of a nonpolar substance like oxygen, only the much weaker *London* forces exist.

Chapter 14   Solutions

2.   A heterogeneous mixture does not have a uniform composition: the composition varies in different places within the mixture. Examples of nonhomogeneous mixtures include salad dressing (mixture of oil, vinegar, water, herbs, and spices) and granite (combination of minerals).

4.   solid

6.   One substance will mix with and dissolve in another substance if the intermolecular forces are similar in the two substances, so that when the mixture forms, the forces between particles in the mixture will be similar to the forces present in the separate substances. Sugar and ethyl alcohol molecules both contain polar -OH groups, which are comparable to the polar -OH structure in water. Sugar or ethyl alcohol molecules can hydrogen bond with water molecules and intermingle with them freely to form a solution. Substances like petroleum (whose molecules contain only carbon and hydrogen) are nonpolar and cannot form interactions with polar water molecules.

8.   independently

10.   unsaturated

12.   large

14.   100.0

16.   a. $\dfrac{2.51 \text{ g KNO}_3}{(47.49 \text{ g H}_2\text{O} + 2.51 \text{ g KNO}_3)} \times 100 = 5.02\% \text{ KNO}_3$

     b. $\dfrac{4.22 \text{ kg KNO}_3}{(51.43 \text{ g H}_2\text{O} + 4.22 \text{ kg KNO}_3)} \times 100 = 7.58\% \text{ KNO}_3$

     c. $\dfrac{6.81 \text{ mg KNO}_3}{(36.39 \text{ mg H}_2\text{O} + 6.81 \text{ mg KNO}_3)} \times 100 = 15.8\% \text{ KNO}_3$

     d. $\dfrac{0.124 \text{ g KNO}_3}{(38.88 \text{ g H}_2\text{O} + 0.124 \text{ g KNO}_3)} \times 100 = 0.318\% \text{ KNO}_3$

18.   To say a solution is 5.25% glucose by mass means that 100. g of the solution would contain 5.25 g of glucose.

     a. $225 \text{ g solution} \times \dfrac{5.25 \text{ g glucose}}{100. \text{ g solution}} = 11.8 \text{ g glucose}$

     b. $225 \text{ g solution} \times \dfrac{5.25 \text{ g glucose}}{100. \text{ g solution}} = 11.8 \text{ g glucose}$

     c. $35 \text{ g glucose} \times \dfrac{100. \text{ g solution}}{5.25 \text{ g glucose}} = 667 \text{ g solution needed}$

d.   225 g solution × $\dfrac{(100. - 5.25) \text{ g water}}{100. \text{ g solution}}$ = 213 g water

Note that the calculation for parts (c) and (d) are identical. These questions are just different ways of asking the same thing: How many g of NaCl are in 1000. g of a 15.0% NaCl solution?

20.   %Cu = $\dfrac{5.31 \text{ g Cu}}{(5.31 \text{ g Cu} + 4.03 \text{ g Zn} + 145 \text{ g Fe})}$ × 100 = $\dfrac{5.31 \text{ g}}{154.3 \text{ g}}$ × 100 = 3.44% Cu

% Zn = $\dfrac{4.03 \text{ g Zn}}{(5.31 \text{ g Cu} + 4.03 \text{ g Zn} + 145 \text{ g Fe})}$ × 100 = $\dfrac{4.03 \text{ g}}{154.3 \text{ g}}$ × 100 = 2.61% Zn

% Fe = $\dfrac{145 \text{ g Fe}}{(5.31 \text{ g Cu} + 4.03 \text{ g Zn} + 145 \text{ g Fe})}$ × 100 = $\dfrac{145 \text{ g}}{154.3 \text{ g}}$ × 100 = 94.0% Fe

22. $\dfrac{67.1 \text{ g } CaCl_2}{(67.1 \text{ g } CaCl_2 + 275 \text{ g } H_2O)}$ × 100 = 19.6% $CaCl_2$

24.   To say that the solution is 6.25% KBr by mass, means that 100. g of the solution will contain 6.25 g KBr.

125 g solution × $\dfrac{6.25 \text{ g KBr}}{100. \text{ g solution}}$ = 7.81 g KBr

26.   To say that the solution is to be 1.25% $CuCl_2$ by mass, means that 100. g of the solution will contain 1.25 g $CuCl_2$

1250. g solution × $\dfrac{1.25 \text{ g } CuCl_2}{100. \text{ g solution}}$ = 15.6 g $CuCl_2$

28.   g heptane = 93 g solution × $\dfrac{}{100. \text{ g solution}}$ = 4.8 g heptane

g pentane = 93 g solution × $\dfrac{2.9 \text{ g pentane}}{100. \text{ g solution}}$ = 2.7 g pentane

g hexane = 93 g solution − 4.8 g heptane − 2.7 g pentane = 86 g hexane

30.   If the solution is 0.121 *M* in $AlCl_3$, then each liter contains 0.121 mol of Al^{3+} ion and 3 × 0.121 = 0.363 mol of Cl^- ion.

32.   To say that a solution has a concentration of 5 *M* means that in 1 L of solution (*not* solvent) there would be 5 mol of solute. To prepare such a solution one would place 5 mol of NaCl in a 1 L flask, and then add whatever amount of water is necessary so that the *total* volume would be 1 L after mixing. The NaCl will occupy some space, so the amount of water to be added will be *less* than 1.00 L.

34. $\text{molarity} = \dfrac{\text{moles of solute}}{\text{liters of solution}}$

a. 250 mL = 0.25 L

$$M = \dfrac{0.50 \text{ mol KBr}}{0.25 \text{ L solution}} = 2.0 \text{ } M$$

b. 500 mL = 0.500 L

$$M = \dfrac{0.50 \text{ mol KBr}}{0.500 \text{ L solution}} = 1.0 \text{ } M$$

c. 750 mL = 0.75 L

$$M = \dfrac{0.50 \text{ mol KBr}}{0.75 \text{ L solution}} = 0.67 \text{ } M$$

d. $M = \dfrac{0.50 \text{ mol KBr}}{1.0 \text{ L solution}} = 0.50 \text{ } M$

36. $\text{molarity} = \dfrac{\text{moles of solute}}{\text{liters of solution}}$

molar mass of KNO_3 = 101.11 g

a. $1.25 \text{ g } KNO_3 \times \dfrac{1 \text{ mol}}{101.11 \text{ g}} = 0.01236 \text{ mol } KNO_3$

$$M = \dfrac{0.01236 \text{ mol } KNO_3}{0.115 \text{ L solution}} = 0.108 \text{ } M$$

b. $12.5 \text{ g } KNO_3 \times \dfrac{1 \text{ mol}}{101.11 \text{ g}} = 0.1236 \text{ mol } KNO_3$

$$M = \dfrac{0.1236 \text{ mol } KNO_3}{1.15 \text{ L solution}} = 0.108 \text{ } M$$

c. $1.25 \text{ mg } KNO_3 \times \dfrac{1 \text{ mmol}}{101.11 \text{ mg}} = 0.01236 \text{ mmol } KNO_3$

$$M = \dfrac{0.01236 \text{ mmol } KNO_3}{1.15 \text{ mL solution}} = 0.0108 \text{ } M$$

d. $1.25 \text{ kg } KNO_3 \times \dfrac{1000 \text{ g}}{1 \text{ kg}} \times \dfrac{1 \text{ mol}}{101.11 \text{ g}} = 12.36 \text{ mol } KNO_3$

$$M = \dfrac{12.36 \text{ mol } KNO_3}{115 \text{ L solution}} = 0.108 \text{ } M$$

38.   molar mass of NaCl = 58.44 g

$$2.71 \text{ g NaCl} \times \frac{1 \text{ mol}}{58.44 \text{ g}} = 0.04637 \text{ mol NaCl}$$

100. mL = 0.100 L

$$M = \frac{0.04637 \text{ mol NaCl}}{0.100 \text{ L solution}} = 0.464 \text{ } M$$

40.   molar mass of I_2 = 253.8 g

225 mL = 0.225 L

$$5.15 \text{ g } I_2 \times \frac{1 \text{ mol}}{253.8 \text{ g}} = 0.0203 \text{ mol } I_2$$

$$M = \frac{0.0203 \text{ mol } I_2}{0.225 \text{ L solution}} = 0.0902 \text{ } M$$

42.   a.   molar mass of NaOH = 40.00 g

$$495 \text{ g NaOH} \times \frac{1 \text{ mol}}{40.00 \text{ g}} = 12.4 \text{ mol NaOH}$$

$$M = \frac{12.4 \text{ mol NaOH}}{20.0 \text{ L solution}} = 0.619 \text{ } M$$

44.   a.   molar mass of HNO_3 = 63.02 g   127 mL = 0.127 L

$$0.127 \text{ L solution} \times \frac{0.105 \text{ mol } HNO_3}{1.00 \text{ L solution}} = 0.0133 \text{ mol } HNO_3$$

$$0.0133 \text{ mol } HNO_3 \times \frac{63.02 \text{ g } HNO_3}{1 \text{ mol } HNO_3} = 0.838 \text{ g } HNO_3$$

b.   molar mass of NH_3 = 17.03 g   155 mL = 0.155 L

$$0.155 \text{ L solution} \times \frac{15.1 \text{ mol } NH_3}{1.00 \text{ L solution}} = 2.34 \text{ mol } NH_3$$

$$2.34 \text{ mol } NH_3 \times \frac{17.03 \text{ g } NH_3}{1 \text{ mol } NH_3} = 39.9 \text{ g } NH_3$$

c.   molar mass KSCN = 97.19 g

$$2.51 \text{ L solution} \times \frac{2.01 \times 10^{-3} \text{ mol KSCN}}{1.00 \text{ L solution}} = 5.05 \times 10^{-3} \text{ mol KSCN}$$

$$5.05 \times 10^{-3} \text{ mol KSCN} \times \frac{97.19 \text{ g KSCN}}{1 \text{ mol KSCN}} = 0.490 \text{ g KSCN}$$

d.   molar mass of HCl = 36.46 g   12.2 mL = 0.0122 L

$$0.0122 \text{ L solution} \times \frac{2.45 \text{ mol HCl}}{1.00 \text{ L solution}} = 0.0299 \text{ mol HCl}$$

$$0.0299 \text{ mol HCl} \times \frac{36.46 \text{ g HCl}}{1 \text{ mol HCl}} = 1.09 \text{ g HCl}$$

46. a.   molar mass of KBr = 119.0 g   173 mL = 0.173 L

$$0.173 \text{ L solution} \times \frac{1.24 \text{ mol KBr}}{1.00 \text{ L solution}} = 0.215 \text{ mol KBr}$$

$$0.215 \text{ mol KBr} \times \frac{119.0 \text{ g}}{1 \text{ mol}} = 25.6 \text{ g KBr}$$

b.   molar mass of HCl = 36.46 g

$$2.04 \text{ L solution} \times \frac{12.1 \text{ mol HCl}}{1.00 \text{ L solution}} = 24.7 \text{ mol HCl}$$

$$24.7 \text{ mol HCl} \times \frac{36.46 \text{ g}}{1 \text{ mol}} = 901 \text{ g HCl}$$

c.   molar mass of NH_3 = 17.03 g   25 mL = 0.025 L

$$0.025 \text{ L solution} \times \frac{3.0 \text{ mol NH}_3}{1.00 \text{ L solution}} = 0.075 \text{ mol NH}_3$$

$$0.075 \text{ mol NH}_3 \times \frac{17.03 \text{ g}}{1 \text{ mol}} = 1.3 \text{ g NH}_3$$

d.   molar mass $CaCl_2$ = 111.0 g   125 mL = 0.125 L

$$0.125 \text{ L solution} \times \frac{0.552 \text{ mol CaCl}_2}{1.00 \text{ L solution}} = 0.0690 \text{ mol CaCl}_2$$

$$0.0690 \text{ mol CaCl}_2 \times \frac{111.0 \text{ g}}{1 \text{ mol}} = 7.66 \text{ g CaCl}_2$$

48. Molar mass of NaCl = 58.44 g   1 lb = 453.59 g

$$453.59 \text{ g NaCl} \times \frac{1 \text{ mol}}{58.44 \text{ g}} = 7.76 \text{ mol NaCl}$$

$$7.76 \text{ mol NaCl} \times \frac{1.00 \text{ L solution}}{1.0 \text{ mol NaCl}} = 7.76 \text{ L} = 7.8 \text{ L of solution}$$

50. a.   10.2 mL = 0.0102 L

$$0.0102 \text{ L} \times \frac{0.451 \text{ mol AlCl}_3}{1.00 \text{ L}} \times \frac{1 \text{ mol Al}^{3+}}{1 \text{ mol AlCl}_3} = 4.60 \times 10^{-3} \text{ mol Al}^{3+}$$

$$0.0102 \text{ L} \times \frac{0.451 \text{ mol AlCl}_3}{1.00 \text{ L}} \times \frac{3 \text{ mol Cl}^-}{1 \text{ mol AlCl}_3} = 1.38 \times 10^{-2} \text{ mol Cl}^-$$

b. $5.51 \text{ L} \times \dfrac{0.103 \text{ mol } Na_3PO_4}{1.00 \text{ L}} \times \dfrac{3 \text{ mol } Na^+}{1 \text{ mol } Na_3PO_4} = 1.70 \text{ mol } Na^+$

   $5.51 \text{ L} \times \dfrac{0.103 \text{ mol } Na_3PO_4}{1.00 \text{ L}} \times \dfrac{1 \text{ mol } PO_4^{3-}}{1 \text{ mol } Na_3PO_4} = 0.568 \text{ mol } PO_4^{3-}$

c. $1.75 \text{ mL} = 0.00175 \text{ L}$

   $0.00175 \text{ L} \times \dfrac{1.25 \text{ mol } CuCl_2}{1.00 \text{ L}} \times \dfrac{1 \text{ mol } Cu^{2+}}{1 \text{ mol } CuCl_2} = 2.19 \times 10^{-3} \text{ mol } Cu^{2+}$

   $0.00175 \text{ L} \times \dfrac{1.25 \text{ mol } CuCl_2}{1.00 \text{ L}} \times \dfrac{2 \text{ mol } Cl^-}{1 \text{ mol } CuCl_2} = 4.38 \times 10^{-3} \text{ mol } Cl^-$

d. $25.2 \text{ mL} = 0.0252 \text{ L}$

   $0.0252 \text{ L} \times \dfrac{0.00157 \text{ mol } Ca(OH)_2}{1.00 \text{ L}} \times \dfrac{1 \text{ mol } Ca^{2+}}{1 \text{ mol } Ca(OH)_2} = 3.96 \times 10^{-5} \text{ mol } Ca^{2+}$

   $0.0252 \text{ L} \times \dfrac{0.00157 \text{ mol } Ca(OH)_2}{1.00 \text{ L}} \times \dfrac{2 \text{ mol } OH^-}{1 \text{ mol } Ca(OH)_2} = 7.91 \times 10^{-5} \text{ mol } OH^-$

52.   Molar mass of $Na_2CO_3 = 106.0 \text{ g}$ $250 \text{ mL} = 0.250 \text{ L}$

   $0.250 \text{ L} \times \dfrac{0.0500 \text{ mol } Na_2CO_3}{1.00 \text{ L}} \times \dfrac{106.0 \text{ g}}{1 \text{ mol}} = 1.33 \text{ g } Na_2CO_3$

54.   half

56. $M_1 \times V_1 = M_2 \times V_2$

a. $M_1 = 0.251 \ M$ $M_2 = ?$

   $V_1 = 125 \text{ mL}$ $V_2 = 250. + 125 = 375 \text{ mL}$

   $M_2 = \dfrac{(0.251 \ M)(125 \text{ mL})}{(375 \text{ mL})} = 0.0837 \ M$

b. $M_1 = 0.499 \ M$ $M_2 = ?$

   $V_1 = 445 \text{ mL}$ $V_2 = 445 + 250. = 695 \text{ mL}$

   $M_2 = \dfrac{(0.499 \ M)(445 \text{ mL})}{(695 \text{ mL})} = 0.320 \ M$

c. $M_1 = 0.101 \ M$ $M_2 = ?$

   $V_1 = 5.25 \text{ L}$ $V_2 = 5.25 + 0.250 = 5.50 \text{ L}$

   $M_2 = \dfrac{(0.101 \ M)(5.25 \text{ L})}{(5.50 \text{ L})} = 0.0964 \ M$

d. $M_1 = 14.5 \ M$ $M_2 = ?$

   $V_1 = 11.2 \text{ mL}$ $V_2 = 11.2 + 250. = 261.2 \text{ mL}$

$$M_2 = \frac{(14.5\ M)\ (11.2\ mL)}{(261.2\ mL)} = 0.622\ M$$

58. $M_1 = 19.4\ M$ $M_2 = 3.00\ M$

   $V_1 = ?\ mL$ $V_2 = 3.50\ L$

$$M_1 = \frac{(3.00\ M)\ (3.50\ mL)}{(19.4\ M)} = 0.541\ L\ (541\ mL)$$

60. $M_1 = 0.211\ M$ $M_2 = ?$

   $V_1 = 75\ mL$ $V_2 = 125\ mL$

$$M_2 = \frac{(0.211\ M)\ (75\ mL)}{(125\ mL)} = 0.127\ M = 0.13\ M$$

62. $M_1 = 0.104\ M$ $M_2 = ?\ M$

   $V_1 = 25.0\ mL$ $V_2 = 100.\ mL$

$$M_2 = \frac{(0.104\ M)\ (25.0\ mL)}{(100.\ mL)} = 0.0260\ M$$

So 54.0 - 25.0 = 29.0 mL of water should be added.

64. $Ba(NO_3)_2(aq) + Na_2SO_4(aq) \rightarrow BaSO_4(s) + 2NaNO_3(aq)$

   12.5 mL = 0.0125 L

$$\text{moles } Ba(NO_3)_2 = 0.0125\ L \times \frac{0.15\ mol\ Ba(NO_3)_2}{1.00\ L} = 1.88 \times 10^{-3}\ mol\ Ba(NO_3)_2$$

From the balanced chemical equation for the reaction, if 1.88×10^{-3} mol $Ba(NO_3)_2$ are to be precipitated, then 1.88×10^{-3} mol Na_2SO_4 will be needed.

$$1.88 \times 10^{-3}\ mol\ Na_2SO_4 \times \frac{1.00\ L}{0.25\ mol\ Na_2SO_4} = 0.0075\ L\ required = 7.5\ mL$$

66. molar mass $Na_2C_2O_4 = 134.0$ g   37.5 mL = 0.0375 L

$$\text{moles } Ca^{2+}\ ion = 0.0375\ L \times \frac{0.104\ mol}{1.00\ L} = 0.00390\ mol\ Ca^{2+}\ ion$$

$$Ca^{2+}(aq) + C_2O_4^{2-}(aq) \rightarrow CaC_2O_4(s)$$

As the precipitation reaction is of 1:1 stoichiometry, then 0.00390 mol of $C_2O_4^{2-}$ ion is needed. Moreover, each formula unit of $Na_2C_2O_4$ contains one $C_2O_4^{2-}$ ion, so 0.00390 mol of $Na_2C_2O_4$ is required.

$$0.00390\ mol\ Na_2C_2O_4 \times \frac{134.0\ g}{1\ mol} = 0.523\ g\ Na_2C_2O_4\ required$$

68.   10.0 mL = 0.0100 L

$$0.0100 \text{ L} \times \frac{0.250 \text{ mol AlCl}_3}{1.00 \text{ L}} = 2.50 \times 10^{-3} \text{ mol AlCl}_3$$

$AlCl_3(aq) + 3NaOH(s) \rightarrow Al(OH)_3(s) + 3NaCl(aq)$

$$2.50 \times 10^{-3} \text{ mol AlCl}_3 \times \frac{3 \text{ mol NaOH}}{1 \text{ mol AlCl}_3} = 7.50 \times 10^{-3} \text{ mol NaOH}$$

molar mass NaOH = 40.0 g

$$7.50 \times 10^{-3} \text{ mol NaOH} \times \frac{40.0 \text{ g NaOH}}{1 \text{ mol}} = 0.300 \text{ g NaOH}$$

70. $HCl(aq) + NaOH(aq) \rightarrow NaCl(aq) + H_2O(l)$

24.9 mL = 0.0249 L

$$0.0249 \text{ L} \times \frac{0.451 \text{ mol NaOH}}{1.00 \text{ L}} = 0.0112 \text{ mol NaOH}$$

$$0.0112 \text{ mol NaOH} \times \frac{1 \text{ mol HCl}}{1 \text{ mol NaOH}} = 0.0112 \text{ mol HCl}$$

$$0.0112 \text{ mol HCl} \times \frac{1.00 \text{ L}}{0.175 \text{ mol HCl}} = 0.0642 \text{ L required} = 64.2 \text{ mL}$$

72.   7.2 mL = 0.0072 L

$$0.0072 \text{ L} \times \frac{2.5 \times 10^{-3} \text{ mol NaOH}}{1.00 \text{ L}} = 1.8 \times 10^{-5} \text{ mol NaOH}$$

$H^+(aq) + OH^-(aq) \rightarrow H_2O(l)$

$$1.8 \times 10^{-5} \text{ mol OH}^- \times \frac{1 \text{ mol H}^+}{1 \text{ mol OH}^-} = 1.8 \times 10^{-5} \text{ mol H}^+$$

100 mL = 0.100 L

$$M = \frac{1.8 \times 10^{-5} \text{ mol H}^+}{0.100 \text{ L}} = 1.8 \times 10^{-4} \text{ } M \text{ H}^+(aq)$$

74.   Experimentally, neutralization reactions are usually performed with volumetric glassware that is calibrated in milliliters rather than liters. For convenience in calculations for such reactions, the arithmetic is often performed in terms of *milli*liters and *milli*moles, rather than in liters and moles: 1 mmol = 0.001 mol. Note that the number of moles of solute per liter of solution, the molarity, is numerically equivalent to the number of *milli*moles of solute per *milli*liter of solution.

a. $HNO_3(aq) + NaOH(aq) \rightarrow NaNO_3(aq) + H_2O(l)$

$$12.7 \text{ mL} \times \frac{0.501 \text{ mmol}}{1 \text{ mL}} = 6.36 \text{ mmol NaOH present in the sample}$$

$$6.36 \text{ mmol NaOH} \times \frac{1 \text{ mmol HNO}_3}{1 \text{ mmol NaOH}} = 6.36 \text{ mmol HNO}_3 \text{ required to react}$$

$$6.36 \text{ mmol HNO}_3 \times \frac{1.00 \text{ mL}}{0.101 \text{ mmol HNO}_3} = 63.0 \text{ mL HNO}_3 \text{ required}$$

b. $2HNO_3(aq) + Ba(OH)_2 \rightarrow Ba(NO_3)_2 + 2H_2O(l)$

$$24.9 \text{ mL} \times \frac{0.00491 \text{ mmol}}{1.00 \text{ mL}} = 0.122 \text{ mmol Ba(OH)}_2 \text{ present in the sample}$$

$$0.122 \text{ mmol Ba(OH)}_2 \times \frac{2 \text{ mmol HNO}_3}{1 \text{ mmol Ba(OH)}_2} = 0.244 \text{ mmol HNO}_3 \text{ required}$$

$$0.244 \text{ mmol HNO}_3 \times \frac{1.00 \text{ mL}}{0.101 \text{ mmol HNO}_3} = 2.42 \text{ mL HNO}_3 \text{ is required}$$

c. $HNO_3(aq) + NH_3(aq) \rightarrow NH_4NO_3(aq)$

$$49.1 \text{ mL} \times \frac{0.103 \text{ mmol}}{1.00 \text{ mL}} = 5.06 \text{ mmol NH}_3 \text{ present in the sample}$$

$$5.06 \text{ mmol NH}_3 \times \frac{1 \text{ mmol HNO}_3}{1 \text{ mmol NH}_3} = 5.06 \text{ mmol HNO}_3 \text{ required}$$

$$5.06 \text{ mmol HNO}_3 \times \frac{1.00 \text{ mL}}{0.101 \text{ mmol HNO}_3} = 50.1 \text{ mL HNO}_3 \text{ required}$$

d. $KOH(aq) + HNO_3(aq) \rightarrow KNO_3(aq) + H_2O(l)$

$$1.21 \text{ L} \times \frac{0.102 \text{ mol}}{1.00 \text{ L}} = 0.123 \text{ mol KOH present in the sample}$$

$$0.123 \text{ mol KOH} \times \frac{1 \text{ mol HNO}_3}{1 \text{ mol KOH}} = 0.123 \text{ mol HNO}_3 \text{ required}$$

$$0.123 \text{ mol HNO}_3 \times \frac{1.00 \text{ L}}{0.101 \text{ mol HNO}_3} = 1.22 \text{ L HNO}_3 \text{ required}$$

76. 1 normal

78. 1.53 equivalents OH^- ion are needed to react with 1.53 equivalents of H^+ ion. By *definition*, one equivalent of OH^- ion exactly neutralizes one equivalent of H^+ ion.

80. $N = \dfrac{\text{number of equivalents of solute}}{\text{number of liters of solution}}$

a. equivalent weight NaOH = molar mass NaOH = 40.00 g

$$0.113 \text{ g NaOH} \times \frac{1 \text{ equiv NaOH}}{40.00 \text{ g}} = 2.83 \times 10^{-3} \text{ equiv NaOH}$$

$$10.2 \text{ mL} = 0.0102 \text{ L}$$

$$N = \frac{2.83 \times 10^{-3} \text{ equiv}}{0.0102 \text{ L}} = 0.277 \ N$$

b.   equivalent weight $Ca(OH)_2 = \dfrac{\text{molar mass}}{2} = \dfrac{74.10 \text{ g}}{2} = 37.05 \text{ g}$

$$12.5 \text{ mg} \times \frac{1 \text{ g}}{10^3 \text{ mg}} \times \frac{1 \text{ equiv}}{37.05 \text{ g}} = 3.37 \times 10^{-4} \text{ equiv } Ca(OH)_2$$

$$100. \text{ mL} = 0.100 \text{ L}$$

$$N = \frac{3.37 \times 10^{-3} \text{ equiv}}{0.100 \text{ L}} = 3.37 \times 10^{-3} \ N$$

c.   equivalent weight $H_2SO_4 = \dfrac{\text{molar mass}}{2} = \dfrac{98.09 \text{ g}}{2} = 49.05 \text{ g}$

$$12.4 \text{ g} \times \frac{1 \text{ equiv}}{49.05 \text{ g}} = 0.253 \text{ equiv } H_2SO_4$$

$$155 \text{ mL} = 0.155 \text{ L}$$

$$N = \frac{0.253 \text{ equiv}}{0.155 \text{ L}} = 1.63 \ N$$

82. a. $0.134 \ M \text{ NaOH} \times \dfrac{1 \text{ equiv NaOH}}{1 \text{ mol NaOH}} = 0.134 \ N \text{ NaOH}$

   b. $0.00521 \ M \ Ca(OH)_2 \times \dfrac{2 \text{ equiv } Ca(OH)_2}{1 \text{ mol } Ca(OH)_2} = 0.0104 \ N \ Ca(OH)_2$

   c. $4.42 \ M \ H_3PO_4 \times \dfrac{3 \text{ equiv } H_3PO_4}{1 \text{ mol } H_3PO_4} = 13.3 \ N \ H_3PO_4$

84. Molar mass of $Ca(OH)_2 = 74.10 \text{ g}$

$$5.21 \text{ mg } Ca(OH)_2 \times \frac{1 \text{ g}}{10^3 \text{ mg}} \times \frac{1 \text{ mol}}{74.10 \text{ g}} = 7.03 \times 10^{-5} \text{ mol } Ca(OH)_2$$

1000. mL = 1.000 L (volumetric flask volume: 4 significant figures).

$$M = \frac{7.03 \times 10^{-5} \text{ mol}}{1.000 \text{ L}} = 7.03 \times 10^{-5} \ M \ Ca(OH)_2$$

$$N = 7.03 \times 10^{-5} \ M \ Ca(OH)_2 \times \frac{2 \text{ equiv } Ca(OH)_2}{1 \text{ mol } Ca(OH)_2} = 1.41 \times 10^{-4} \ N \ Ca(OH)_2$$

86. $H_3PO_4(aq) + 3NaOH \rightarrow Na_3PO_4(aq) + 3H_2O(l)$

   $0.204 \ M \text{ NaOH} = 0.204 \ N \text{ NaOH}$ $18.7 \text{ mL} = 0.0187 \text{ L}$

$$0.0187 \text{ L NaOH} \times \frac{0.204 \text{ equiv}}{1.00 \text{ L}} = 0.00381 \text{ equiv NaOH}$$

0.00381 equiv NaOH requires 0.00381 equiv H_3PO_4 to react.

$$0.00381 \text{ equiv } H_3PO_4 \times \frac{1.00 \text{ L}}{0.105 \text{ equiv}} = 0.0363 \text{ L} = 36.3 \text{ mL } H_3PO_4 \text{ solution}$$

88. $2NaOH(aq) + H_2SO_4(aq) \rightarrow Na_2SO_4(aq) + 2 H_2O(l)$

$$27.34 \text{ mL NaOH} \times \frac{0.1021 \text{ mmol}}{1.00 \text{ mL}} = 2.791 \text{ mmol NaOH}$$

$$2.791 \text{ mmol NaOH} \times \frac{1 \text{ mmol } H_2SO_4}{2 \text{ mmol NaOH}} = 1.396 \text{ mmol } H_2SO_4$$

$$M = \frac{1.396 \text{ mmol } H_2SO_4}{25.00 \text{ mL}} = 0.05583 \ M \ H_2SO_4 = 0.1117 \ N \ H_2SO_4$$

90. Molarity is defined as the number of moles of solute contained in 1 liter of *total* solution volume (solute plus solvent after mixing). In the first case, where 50. g of NaCl is dissolved in 1.0 L of water, the total volume after mixing is *not* known and the molarity cannot be calculated. In the second example, the final volume after mixing is known and the molarity can be calculated simply.

92. $$75 \text{ g solution} \times \frac{25 \text{ g NaCl}}{100 \text{ g solution}} = 18.75 \text{ g NaCl}$$

$$\text{new \%} = \frac{18.75 \text{ g NaCl}}{575 \text{ g solution}} \times 100 = 3.26 = 3.3 \ \%$$

94. $Ba(NO_3)_2(aq) + H_2SO_4(aq) \rightarrow BaSO_4(s) + 2HNO_3(aq)$

37.5 mL = 0.0375 L

$$0.0375 \text{ L} \times \frac{0.221 \text{ mol } H_2SO_4}{1.00 \text{ L}} = 0.00829 \text{ mol } H_2SO_4$$

As the coefficients of $Ba(NO_3)_2$ and H_2SO_4 in the balanced chemical equation for the reaction are both *one*, then 0.00829 mol of Ba^{2+} ion will be precipitated from the solution as $BaSO_4$.

molar mass $BaSO_4$ = 233.4 g

$$0.00829 \text{ mol } BaSO_4 \times \frac{233.4 \text{ g } BaSO_4}{1 \text{ mol } BaSO_4} = 1.93 \text{ g } BaSO_4 \text{ precipitate}$$

96. molar mass H_2O = 18.0 g

1.0 L water = 1.0×10^3 mL water $\approx 1.0 \times 10^3$ g water

$$1.0 \times 10^3 \text{ g } H_2O \times \frac{1 \text{ mol } H_2O}{18.0 \text{ g } H_2O} = 56 \text{ mol } H_2O$$

98.   500 mL HCl solution = 0.500 L HCl solution

$$0.500 \text{ L solution} \times \frac{0.100 \text{ mol HCl}}{1.00 \text{ L HCl solution}} = 0.0500 \text{ mol HCl}$$

$$0.0500 \text{ mol HCl} \times \frac{22.4 \text{ L HCl gas at STP}}{1 \text{ mol HCl}} = 1.12 \text{ L HCl gas at STP}$$

100. $$10.0 \text{ g HCl} \times \frac{100. \text{ g solution}}{33.1 \text{ g HCl}} = 30.21 \text{ g solution}$$

$$30.21 \text{ g solution} \times \frac{1.00 \text{ mL solution}}{1.147 \text{ g solution}} = 26.3 \text{ mL solution}$$

102.   molar mass $CaCl_2$ = 111.0 g

$$14.2 \text{ g CaCl}_2 \times \frac{1 \text{ mol CaCl}_2}{111.0 \text{ g CaCl}_2} = 0.128 \text{ mol CaCl}_2$$

50.0 mL = 0.0500 L

$$M = \frac{0.128 \text{ mol CaCl}_2}{0.0500 \text{ L}} = 2.56 \ M$$

104.   a. $$\frac{5.0 \text{ g KNO}_3}{(5.0 \text{ g KNO}_3 + 75 \text{ g H}_2\text{O})} \times 100 = \frac{5.0 \text{ g}}{80.0 \text{ g}} \times 100 = 6.3\% \text{ KNO}_3$$

b.   2.5 mg = 0.0025 g

$$\frac{0.0025 \text{ g KNO}_3}{(0.0025 \text{ g KNO}_3 + 1.0 \text{ g H}_2\text{O})} \times 100 = \frac{0.0025 \text{ g}}{1.0025 \text{ g}} \times 100 = 0.25\% \text{ KNO}_3$$

c. $$\frac{11 \text{ g KNO}_3}{(11 \text{ g KNO}_3 + 89 \text{ g H}_2\text{O})} \times 100 = \frac{11 \text{ g}}{100 \text{ g}} \times 100 = 11\% \text{ KNO}_3$$

d. $$\frac{11 \text{ g KNO}_3}{(11 \text{ g KNO}_3 + 49 \text{ g H}_2\text{O})} \times 100 = \frac{11 \text{ g}}{60 \text{ g}} \times 100 = 18\% \text{ KNO}_3$$

106. $$\%\text{C} = \frac{5.0 \text{ g C}}{(5.0 \text{ g C} + 1.5 \text{ g Ni} + 100. \text{ g Fe})} \times 100 = \frac{5.0 \text{ g}}{106.5 \text{ g}} \times 100 = 4.7\% \text{ C}$$

$$\%\text{Ni} = \frac{1.5 \text{ g Ni}}{(5.0 \text{ g C} + 1.5 \text{ g Ni} + 100. \text{ g Fe})} \times 100 = \frac{1.5 \text{ g}}{106.5 \text{ g}} \times 100 = 1.4\% \text{ Ni}$$

$$\%\text{Fe} = \frac{100. \text{ g Fe}}{(5.0 \text{ g C} + 1.5 \text{ g Ni} + 100. \text{ g Fe})} \times 100 = \frac{100. \text{ g}}{106.5 \text{ g}} \times 100 = 93.9\% \text{ Fe}$$

108. To say that the solution is 5.5% by mass Na_2CO_3 means that 5.5 g of Na_2CO_3 are contained in every 100 g of the solution.

$$500.\ g\ solution \times \frac{5.5\ g\ Na_2CO_3}{100\ g\ solution} = 28\ g\ Na_2CO_3$$

110. For NaCl: $125\ g\ solution \times \dfrac{7.5\ g\ NaCl}{100\ g\ solution} = 9.4\ g\ NaCl$

   For KBr: $125\ g\ solution \times \dfrac{2.5\ g\ KBr}{100\ g\ solution} = 3.1\ g\ KBr$

112. $molarity = \dfrac{moles\ of\ solute}{liters\ of\ solution}$

   a.   25 mL = 0.025 L

   $$M = \frac{0.10\ mol\ CaCl_2}{0.025\ L\ solution} = 4.0\ M$$

   b. $M = \dfrac{2.5\ mol\ KBr}{2.5\ L\ solution} = 1.0\ M$

   c.   755 mL = 0.755 L

   $$M = \frac{0.55\ mol\ NaNO_3}{0.755\ L\ solution} = 0.73\ M$$

   d. $M = \dfrac{4.5\ mol\ Na_2SO_4}{1.25\ L\ solution} = 3.6\ M$

114. molar mass $C_{12}H_{22}O_{11}$ = 342.3 g

   $$125\ g\ C_{12}H_{22}O_{11} \times \frac{1\ mol}{342.3\ g} = 0.3652\ mol\ C_{12}H_{22}O_{11}$$

   450. mL = 0.450 L

   $$M = \frac{0.3652\ mol\ C_{12}H_{22}O_{11}}{0.450\ L\ solution} = 0.812\ M$$

116. molar mass NaCl = 58.44 g

   $$1.5\ g\ NaCl \times \frac{1\ mol}{58.44\ g} = 0.0257\ mol\ NaCl$$

   $$M = \frac{0.0257\ mol\ NaCl}{1.0\ L\ solution} = 0.026\ M$$

118. a. $4.25\ L\ solution \times \dfrac{0.105\ mol\ KCl}{1.00\ L\ solution} = 0.446\ mol\ KCl$

   molar mass KCl = 74.6 g

$$0.446 \text{ mol KCl} \times \frac{74.6 \text{ g KCl}}{1 \text{ mol KCl}} = 33.3 \text{ g KCl}$$

b.   15.1 mL = 0.0151 L

$$0.0151 \text{ L solution} \times \frac{0.225 \text{ mol NaNO}_3}{1.00 \text{ L solution}} = 3.40 \times 10^{-3} \text{ mol NaNO}_3$$

molar mass $NaNO_3$ = 85.00 g

$$3.40 \times 10^{-3} \text{ mol} \times \frac{85.00 \text{ g NaNO}_3}{1 \text{ mol NaNO}_3} = 0.289 \text{ g NaNO}_3$$

c.   25 mL = 0.025 L

$$0.025 \text{ L solution} \times \frac{3.0 \text{ mol HCl}}{1.00 \text{ L solution}} = 0.075 \text{ mol HCl}$$

molar mass HCl = 36.46 g

$$0.075 \text{ mol HCl} \times \frac{36.46 \text{ g HCl}}{1 \text{ mol HCl}} = 2.7 \text{ g HCl}$$

d.   100. mL = 0.100 L

$$0.100 \text{ L solution} \times \frac{0.505 \text{ mol H}_2\text{SO}_4}{1.00 \text{ L solution}} = 0.0505 \text{ mol H}_2\text{SO}_4$$

molar mass H_2SO_4 = 98.09 g

$$0.0505 \text{ mol H}_2\text{SO}_4 \times \frac{98.09 \text{ g H}_2\text{SO}_4}{1 \text{ mol H}_2\text{SO}_4} = 4.95 \text{ g H}_2\text{SO}_4$$

120. a. $$1.25 \text{ L} \times \frac{0.250 \text{ mol Na}_3\text{PO}_4}{1.00 \text{ L}} = 0.3125 \text{ mol Na}_3\text{PO}_4$$

$$0.3125 \text{ mol Na}_3\text{PO}_4 \times \frac{3 \text{ mol Na}^+}{1 \text{ mol Na}_3\text{PO}_4} = 0.938 \text{ mol Na}^+$$

$$0.3125 \text{ mol Na}_3\text{PO}_4 \times \frac{1 \text{ mol PO}_4^{3-}}{1 \text{ mol Na}_3\text{PO}_4} = 0.313 \text{ mol PO}_4^{3-}$$

b.   3.5 mL = 0.0035 L

$$0.0035 \text{ L} \times \frac{6.0 \text{ mol H}_2\text{SO}_4}{1.00 \text{ L}} = 0.021 \text{ mol H}_2\text{SO}_4$$

$$0.021 \text{ mol H}_2\text{SO}_4 \times \frac{2 \text{ mol H}^+}{1 \text{ mol H}_2\text{SO}_4} = 0.042 \text{ mol H}^+$$

$$0.021 \text{ mol H}_2\text{SO}_4 \times \frac{1 \text{ mol SO}_4^{2-}}{1 \text{ mol H}_2\text{SO}_4} = 0.021 \text{ mol SO}_4^{2-}$$

c.   25 mL = 0.025 L

$$0.025 \text{ L} \times \frac{0.15 \text{ mol AlCl}_3}{1.00 \text{ L}} = 0.00375 \text{ mol AlCl}_3$$

$$0.00375 \text{ mol AlCl}_3 \times \frac{1 \text{ mol Al}^{3+}}{1 \text{ mol AlCl}_3} = 0.0038 \text{ mol Al}^{3+}$$

$$0.00375 \text{ mol AlCl}_3 \times \frac{3 \text{ mol Cl}^-}{1 \text{ mol AlCl}_3} = 0.011 \text{ mol Cl}^-$$

d. $$1.50 \text{ L} \times \frac{1.25 \text{ mol BaCl}_2}{1.00 \text{ L}} = 1.875 \text{ mol BaCl}_2$$

$$1.875 \text{ mol BaCl}_2 \times \frac{1 \text{ mol Ba}^{2+}}{1 \text{ mol BaCl}_2} = 1.88 \text{ mol Ba}^{2+}$$

$$1.875 \text{ mol BaCl}_2 \times \frac{2 \text{ mol Cl}^-}{1 \text{ mol BaCl}_2} = 3.75 \text{ mol Cl}^-$$

122. $M_1 \times V_1 = M_2 \times V_2$

a. $M_1 = 0.200 \ M$ $M_2 = ?$

$V_1 = 125 \text{ mL}$ $V_2 = 125 + 150. = 275 \text{ mL}$

$$M_2 = \frac{(0.200 \ M)(125 \text{ mL})}{(275 \text{ mL})} = 0.0909 \ M$$

b. $M_1 = 0.250 \ M$ $M_2 = ?$

$V_1 = 155 \text{ mL}$ $V_2 = 155 + 150. = 305 \text{ mL}$

$$M_2 = \frac{(0.250 \ M)(155 \text{ mL})}{(305 \text{ mL})} = 0.127 \ M$$

c. $M_1 = 0.250 \ M$ $M_2 = ?$

$V_1 = 0.500 \text{ L} = 500. \text{ mL}$ $V_2 = 500. + 150. = 650. \text{ mL}$

$$M_2 = \frac{(0.250 \ M)(500. \text{ mL})}{(650 \text{ mL})} = 0.192 \ M$$

d. $M_1 = 18.0 \ M$ $M_2 = ?$

$V_1 = 15 \text{ mL}$ $V_2 = 15 + 150. = 165 \text{ mL}$

$$M_2 = \frac{(18.0 \ M)(15 \text{ mL})}{(165 \text{ mL})} = 1.6 \ M$$

124. $M_1 \times V_1 = M_2 \times V_2$

$M_1 = 5.4 \ M$ $M_2 = ?$

$V_1 = 50. \text{ mL}$ $V_2 = 300. \text{ mL}$

$$M_2 = \frac{(5.4 \; M)(50. \; mL)}{(300. \; mL)} = 0.90 \; M$$

126.   25.0 mL = 0.0250 L

$$0.0250 \; L \; NiCl_2 \; solution \times \frac{0.20 \; mol \; NiCl_2}{1.00 \; L \; NiCl_2 \; solution} = 0.00500 \; mol \; NiCl_2$$

$$0.00500 \; mol \; NiCl_2 \times \frac{1 \; mol \; Na_2S}{1 \; mol \; NiCl_2} = 0.00500 \; mol \; Na_2S$$

$$0.00500 \; mol \; Na_2S \times \frac{1.00 \; L \; Na_2S \; solution}{0.10 \; mol \; Na_2S} = 0.050 \; L = 50. \; mL \; Na_2S \; solution$$

128. $HNO_3(aq) + NaOH(aq) \rightarrow NaNO_3(aq) + H_2O(l)$

35.0 mL = 0.0350 L

$$0.0350 \; L \times \frac{0.150 \; mol \; NaOH}{1.00 \; L} = 5.25 \times 10^{-3} \; mol \; NaOH$$

$$5.25 \times 10^{-3} \; mol \; NaOH \times \frac{1 \; mol \; HNO_3}{1 \; mol \; NaOH} = 5.25 \times 10^{-3} \; mol \; HNO_3$$

$$5.25 \times 10^{-3} \; mol \; HNO_3 \times \frac{1.00 \; L}{0.150 \; mol \; HNO_3} = 0.0350 \; L = 35.0 \; mL \; HNO_3$$

130. $N = \dfrac{\text{number of equivalents of solute}}{\text{number of liters of solution}}$

a.   equivalent weight HCl = molar mass HCl = 36.46 g

$$15.0 \; g \; HCl \times \frac{1 \; equiv \; HCl}{36.46 \; g \; HCl} = 0.411 \; equiv \; HCl$$

500. mL = 0.500 L

$$N = \frac{0.411 \; equiv}{0.500 \; L} = 0.822 \; N$$

b.   equivalent weight $H_2SO_4 = \dfrac{\text{molar mass}}{2} = \dfrac{98.09 \; g}{2} = 49.05 \; g$

$$49.0 \; g \; H_2SO_4 \times \frac{1 \; equiv \; H_2SO_4}{49.05 \; g \; H_2SO_4} = 0.999 \; equiv \; H_2SO_4$$

250. mL = 0.250 L

$$N = \frac{0.999 \; equiv}{0.250 \; L} = 4.00 \; N$$

c.   equivalent weight $H_3PO_4 = \dfrac{\text{molar mass}}{3} = \dfrac{98.0 \; g}{3} = 32.67 \; g$

$$10.0 \text{ g } H_3PO_4 \times \frac{1 \text{ equiv } H_3PO_4}{32.67 \text{ g } H_3PO_4} = 0.3061 \text{ equiv } H_3PO_4$$

$$100. \text{ mL} = 0.100 \text{ L}$$

$$N = \frac{0.3061 \text{ equiv}}{0.100 \text{ L}} = 3.06 \ N$$

132. molar mass $NaH_2PO_4 = 120.0 \text{ g}$

$$5.0 \text{ g } NaH_2PO_4 \times \frac{1 \text{ mol } NaH_2PO_4}{120.0 \text{ g } NaH_2PO_4} = 0.04167 \text{ mol } NaH_2PO_4$$

$$500. \text{ mL} = 0.500 \text{ L}$$

$$M = \frac{0.04167 \text{ mol}}{0.500 \text{ L}} = 0.08333 \ M \ NaH_2PO_4 = 0.083 \ M \ NaH_2PO_4$$

$$0.08333 \ M \ NaH_2PO_4 \times \frac{2 \text{ equiv } NaH_2PO_4}{1 \text{ mol } NaH_2PO_4} = 0.1667 \ N \ NaH_2PO_4 = 0.17 \ N \ NaH_2PO_4$$

134. $N_{acid} \times V_{acid} = N_{base} \times V_{base}$

$N_{acid} \times (10.0 \text{ mL}) = (3.5 \times 10^{-2} \ N)(27.5 \text{ mL})$

$N_{acid} = 9.6 \times 10^{-2} \ N \ HNO_3$

Cumulative Review: Chapters 12, 13, and 14

2.   The pressure exerted by the atmosphere is due to the several mile thick
      layer of gases above the surface of the earth pressing down on us.
      Atmospheric pressure has traditionally been measured with a mercury
      barometer (see Figure 12.2). A mercury barometer usually consists of a
      glass tube that is sealed at one end and filled with mercury. The tube
      is then inverted over an open reservoir also containing mercury. When
      the tube is inverted, most of the mercury does not fall out of the tube.
      As the reservoir of mercury is open to the atmosphere, the atmospheric
      pressure on the surface of the mercury in the reservoir is of sufficient
      pressure to hold the bulk of the mercury in the glass tube; the pressure
      of the atmosphere is sufficient, on average, to support a column of
      mercury 76 cm (760 mm) high in the tube.

4.   In simple terms, Boyle's law states that the volume of a gas sample will
      decrease if you squeeze harder on it. Imagine squeezing hard on a tennis
      ball with your hand: the ball collapses as the gas inside is forced into
      a smaller volume by your hand. Of course, to be perfectly correct, the
      temperature and amount of gas (moles) must remain the same as you adjust
      the pressure for Boyle's law to hold true. There are two mathematical
      statements of Boyle's law you should remember. The first is

$$P \times V = \text{constant},$$

      which basically is the definition of Boyle's law (in order for the
      product ($P \times V$) to remain constant, if one of these terms increases the
      other must decrease). The second formula is the one more commonly used
      in solving problems,

$$P_1 \times V_1 = P_2 \times V_2.$$

      With this second formulation, we can determine pressure-volume
      information about a given sample under two sets of conditions. These two
      mathematical formulas are just two different ways of saying the same
      thing: if the pressure on a sample of gas is increased, the volume of
      the sample of gas will decrease. A graph of Boyle's law data is given as
      Figure 12.5: this type of graph ($xy = k$) is known to mathematicians as a
      hyperbola.

6.   Charles's law simply says that if you heat a sample of gas, the volume
      of the sample will increase. That is, when the temperature of a gas is
      increased, the volume of the gas also increases (assuming the pressure
      and amount of gas remains the same). Charles's law is a direct
      proportionality when the temperature is expressed in kelvins (if you
      increase T, this increases V), whereas Boyle's law is an inverse
      proportionality (if you increase P, this decreases V). There are two
      mathematical statements of Charles's law with which you should be
      familiar. The first statement is:

$$V = kT.$$

      This is simply a definition (the volume of a gas sample is directly
      related to its Kelvin temperature: if you increase the temperature, the

volume increases). The working formulation of Charles's law we use in problem solving is given as:

$$\frac{V_1}{T_1} = \frac{V_2}{T_2}$$

With this formulation, we can determine volume-temperature information for a given gas sample under two sets of conditions. Charles's law only holds true if the amount of gas remains the same (obviously the volume of a gas sample would increase if there were more gas present) and also if the pressure remains the same (a change in pressure also changes the volume of a gas sample).

8. Avogadro's law tells us that, with all other things being equal, two moles of gas are twice as big as one mole of gas! That is, the volume of a sample of gas is directly proportional to the number of moles or molecules of gas present (at constant temperature and pressure). If we want to compare the volumes of two samples of the same gas as an indication of the amount of gas present in the samples, we would have to make certain that the two samples of gas are at the same pressure and temperature: the volume of a sample of gas would vary with either temperature or pressure, or both. Avogadro's law holds true for comparing gas samples that are under the same conditions. Avogadro's law is a direct proportionality: the greater the number of gas molecules you have in a sample, the larger the sample's volume will be.

10. The "partial" pressure of an individual gas in a mixture of gases represents the pressure the gas would have in the same container at the same temperature if it were the only gas present. The total pressure in a mixture of gases is the sum of the individual partial pressures of the gases present in the mixture. Because the partial pressures of the gases in a mixture are additive (i.e., the total pressure is the sum of the partial pressures), this suggests that the total pressure in a container is a function only of the number of molecules present in the same, and not of the identity of the molecules or any other property of the molecules (such as their inherent atomic size).

12. The main postulates of the kinetic-molecular theory for gases are as follows: (a) gases consist of tiny particles (atoms or molecules), and the size of these particles is negligible compared to the bulk volume of a gas sample; (b) the particles in a gas are in constant random motion, colliding with each other and with the walls of the container; (c) the particles in a gas sample do not exert any attractive or repulsive forces on one another; (d) the average kinetic energy of the particles in a sample of gas is directly related to the absolute temperature of the gas sample. The pressure exerted by a gas is a result of the molecules colliding with (and pushing on) the walls of the container. The pressure increases with temperature because at a higher temperature, the molecules are moving faster and hit the walls of the container with greater force. A gas fills whatever volume is available to it because the molecules in a gas are in constant random motion: if the motion of

the molecules is random, they eventually will move out into whatever volume is available until the distribution of molecules is uniform. At constant pressure, the volume of a gas sample increases as the temperature is increased because with each collision having greater force, the container must expand so that the molecules (and therefore the collisions) are farther apart if the pressure is to remain constant.

14.   Solids and liquids are much more condensed states of matter than are gases: the molecules are much closer together in solids and liquids and interact with each other to a much greater extent. Solids and liquids have much greater densities than do gases, and are much less compressible, because there is so little room between the molecules in the solid and liquid states (solids and liquids effectively have native volumes of their own, and their volumes are not affected nearly as much by the temperature or pressure). Although solids are more rigid than liquids, the solid and liquid state have much more in common with each other than either of these states has with the gaseous state. We know this is true because it typically only takes a few kilojoules of energy to melt 1 mol of a solid (not much change has to take place in the molecules), whereas it may take 10 times more energy to vaporize a liquid (as there is a great change between the liquid and gaseous states).

16.   The normal boiling point of water, that is, water's boiling point at a pressure of exactly 760 mm Hg, is 100°C (you will recall that the boiling point of water was used to set one of the reference temperatures of the Celsius temperature scale). Water remains at 100°C while boiling, until all the water has boiled away, because the additional heat energy being added to the sample is used to overcome attractive forces among the water molecules as they go from the condensed, liquid state to the gaseous state. The normal (760 mm Hg) freezing point of water is exactly 0°C (again, this property of water was used as one of the reference points for the Celsius temperature scale). A cooling curve for water is given in Figure 13.2. Notice how the curve shows that the amount of heat needed to boil the sample is much larger than the amount needed to melt the sample.

18.   Dipole-dipole forces are a type of intermolecular force that can exist between molecules with permanent dipole moments. Molecules with permanent dipole moments try to orient themselves so that the positive end of one polar molecule can attract the negative end of another polar molecule. Dipole-dipole forces are not nearly as strong as ionic or covalent bonding forces (only about 1% as strong as covalent bonding forces) because electrostatic attraction is related to the magnitude of the charges of the attracting species. As polar molecules have only a "partial" charge at each end of the dipole, the magnitude of the attractive force is not as large. The strength of such forces also drops rapidly as molecules become farther apart and is important only in the solid and liquid states (such forces are negligible in the gaseous state because the molecules are too far apart). Hydrogen bonding is an especially strong sort of dipole-dipole attractive force that can exist when hydrogen atoms are directly bonded to the most strongly

electronegative atoms (N, O, and F). Because the hydrogen atom is so small, dipoles involving N-H, O-H, and F-H bonds can approach each other much more closely than can dipoles involving other atoms. As the magnitude of dipole-dipole forces is dependent on distance, unusually strong attractive forces can exist in such molecules. We take the fact that the boiling point of water is higher than that of the other covalent hydrogen compounds of the Group 6 elements as evidence for the special strength of hydrogen bonding (it takes more energy to vaporize water because of the extra strong forces holding together the molecules in the liquid state).

20. Vaporization of a liquid requires an input of energy because the intermolecular forces that hold the molecules together in the liquid state must be overcome. The high heat of vaporization of water is essential to life on Earth because much of the excess energy striking the Earth from the sun is dissipated in vaporizing water. Condensation is the opposite process to vaporization; that is, condensation refers to the process by which molecules in the vapor state form a liquid. In a closed container containing a liquid and some empty space above the liquid, an equilibrium is set up between vaporization and condensation. The liquid in such a sealed container never completely evaporates: when the liquid is first placed in the container, the liquid phase begins to evaporate into the empty space. As the number of molecules in the vapor phase begins to get large, however, some of these molecules begin to re-enter the liquid phase. Eventually, every time a molecule of liquid somewhere in the container enters the vapor phase, somewhere else in the container a molecule of vapor re-enters the liquid. There is no further net change in the amount of liquid phase (although molecules are continually moving between the liquid and vapor phases). The pressure of the vapor in such an equilibrium situation is characteristic for the liquid at each particular temperature (for example, the vapor pressures of water are tabulated at different temperatures in Table 12.2). A simple experiment to determine vapor pressure is shown in Figure 13.10. Samples of a liquid are injected into a sealed tube containing mercury; because mercury is so dense, the liquids float to the top of the mercury where they evaporate. As the vapor pressures of the liquids develop to the saturation point, the level of mercury in the tube changes as an index of the magnitude of the vapor pressures. Typically, liquids with strong intermolecular forces have small vapor pressures (they have more difficulty in evaporating) than do liquids with very weak intermolecular forces: for example, the components of gasoline (weak forces) have much higher vapor pressures, and evaporate more easily than does water (strong forces).

22. The simple model we use to explain many properties of metallic elements is called the electron sea model. In this model we picture a regular lattice array of metal cations in sort of a "sea" of mobile valence electrons. The electrons can move easily to conduct heat or electricity through the metal; and the lattice of cations can be deformed fairly easily, allowing the metal to be hammered into a sheet or stretched to make a wire. An alloy contains a mixture of elements, which overall has metallic properties. Substitutional alloys consist of a host metal in

which some of the atoms in the metal's crystalline structure are replaced by atoms of other metallic elements of comparable size to the atoms of the host metal. For example, sterling silver consists of an alloy in which approximately 7% of the silver atoms have been replaced by copper atoms. Brass and pewter are also substitutional alloys. An interstitial alloy is formed when other smaller atoms enter the interstices (holes) between atoms in the host metal's crystal structure. Steel is an interstitial alloy in which typically carbon atoms enter the interstices of a crystal of iron atoms. The presence of the interstitial carbon atoms markedly changes the properties of the iron, making it much harder, more malleable, and more ductile. Depending on the amount of carbon introduced into the iron crystals, the properties of the resulting steel can be carefully controlled.

24. A saturated solution is one that contains as much solute as can dissolve at a particular temperature. To say that a solution is saturated does not necessarily mean that the solute is present at a high concentration. For example, magnesium hydroxide only dissolves to a very small extent before the solution is saturated, whereas it takes a great deal of sugar to form a saturated solution (and the saturated solution is extremely concentrated). A saturated solution is one which is in equilibrium with undissolved solute: as molecules of solute dissolve from the solid in one place in the solution, dissolved molecules rejoin the solid phase in another place in the solution. As with the development of vapor pressure above a liquid (see Question 20 above), formation of a solution reaches a state of dynamic equilibrium: once the rates of dissolving and "undissolving" become equal, there will be no further net change in the concentration of the solution and the solution will be saturated.

26. Adding more solvent to a solution so as to dilute the solution *does not change* the number of moles of solute present, but only changes the volume in which the solute is dispersed. If we are using the molarity of the solution to describe its concentration, the number of liters is changed when we add solvent, and the number of moles per liter (the molarity) changes, but the actual number of moles of solute does not change. For example, 125 mL of 0.551 M NaCl contains 68.9 millimol of NaCl. The solution will still contain 68.9 millimol of NaCl after the 250 mL of water is added to it, only now the 68.9 millimol of NaCl will be dispersed in a total volume of 375 mL. This gives the new molarity as 68.9 mmol/375 mL = 0.184 M. The volume and the concentration have changed, but the number of moles of solute in the solution has not changed.

28. $P_1 \times V_1 = P_2 \times V_2$

   a. $V_2 = \dfrac{(781 \text{ mm Hg})(275 \text{ mL})}{(1034 \text{ mm Hg})} = 208 \text{ mL}$

   b. $P_2 = \dfrac{(1.02 \text{ atm})(45.3 \text{ L})}{(100. \text{ L})} = 0.462 \text{ atm}$

c.   Let P represent the initial pressure of the gas; the final pressure of the gas will then be $(P/2)$. Then, if the volume of the gas sample is represented by V

$$\text{new volume} = \frac{(P \text{ mm Hg})(V \text{ mL})}{[(P/2) \text{ mm Hg}]} = 2V \text{ mL}.$$

When the pressure is halved, the volume of the gas sample is doubled (as Boyle's law predicts).

30.   a. $P_1 = 1.51$ atm $P_2 = 1.00$ atm

$V_1 = 245$ mL $V_2 = ?$ mL

$T_1 = 29 \text{ °C} = 302$ K $T_2 = 273$ K

$$V_2 = \frac{P_1 V_1 T_2}{T_1 P_2} = \frac{(1.51 \text{ atm})(245 \text{ mL})(273 \text{ K})}{(302 \text{ K})(1.00 \text{ atm})} = 334 \text{ mL}$$

b.   molar mass of $N_2 = 28.02$ g

$$21.6 \text{ g N}_2 \times \frac{1 \text{ mol N}_2}{28.02 \text{ g N}_2} = 0.771 \text{ mol N}_2$$

As one mole of an ideal gas occupies a volume of 22.4 L at STP

$$0.771 \text{ mol N}_2 \times \frac{22.4 \text{ Liters}}{1 \text{ mol N}_2} = 17.3 \text{ L}.$$

c.   molar masses: He, 4.003 g; Ne, 20.18 g

$$1.62 \text{ g He} \times \frac{1 \text{ mol He}}{4.003 \text{ g He}} = 0.4046 \text{ mol He}$$

$$2.41 \text{ g Ne} \times \frac{1 \text{ mol Ne}}{20.18 \text{ g Ne}} = 0.1194 \text{ mol Ne}$$

At STP, the total pressure is 1.00 atm.

$$P_{\text{He}} = \frac{0.4046 \text{ mol He}}{(0.4046 + 0.1194) \text{ mol}} \times 1.00 \text{ atm} = 0.772 \text{ atm}$$

$$P_{\text{Ne}} = \frac{0.1194 \text{ mol Ne}}{(0.4046 + 0.1194) \text{ mol}} \times 1.00 \text{ atm} = 0.228 \text{ atm}$$

d.   molar mass of He, 4.003 g   11.2 mL = 0.0112 L

One mole of an ideal gas has a volume of 22.4 L at STP.

$$0.0112 \text{ L} \times \frac{1 \text{ mol He}}{22.4 \text{ L}} \times \frac{4.003 \text{ g He}}{1 \text{ mol He}} = 0.00200 \text{ g He}$$

32.   As we have 100. g of the solution, which is 3.11% by mass H_2O_2, then the sample must contain 3.11 g of H_2O_2

molar mass H_2O_2 = 34.02 g

$$3.11 \text{ g } H_2O_2 \times \frac{1 \text{ mol } H_2O_2}{34.02 \text{ g } H_2O_2} = 0.0914 \text{ mol } H_2O_2$$

$$0.0914 \text{ mol } H_2O_2 \times \frac{1 \text{ mol } O_2}{2 \text{ mol } H_2O_2} = 0.0457 \text{ mol } O_2 \text{ produced}$$

At STP, $0.0457 \text{ mol } O_2 \times \dfrac{22.4 \text{ L } O_2}{1 \text{ mol } O_2} = 1.02 \text{ L } O_2$

P_1 = 760 mm Hg P_2 = 771 mm Hg

V_1 = 1.02 L V_2 = ?

T_1 = 273 K T_2 = 297 K

$$V_2 = \frac{P_1 V_1 T_2}{T_1 P_2} = \frac{(760 \text{ mm Hg}) (1.02 \text{ L}) (297 \text{ K})}{(273 \text{ K}) (771 \text{ mm Hg})} = 1.09 \text{ L}$$

34.   a.   mass of solution = 1.01 + 2.11 + 1.55 + 151 = 155.67 g

$$\frac{1.01 \text{ g NaCl}}{155.67 \text{ g}} \times 100 = 0.649\% \text{ NaCl}$$

b.   mass of solution (part a) = 155.67 g; molar mass NaCl = 58.44 g

$$\text{mol NaCl} = \text{mol Na}^+ = 1.01 \text{ g NaCl} \times \frac{1 \text{ mol NaCl}}{58.44 \text{ g NaCl}} = 0.01728 \text{ mol}$$

$$\text{volume} = 155.67 \text{ g} \times \frac{1 \text{ mL}}{1.02 \text{ g}} \times \frac{1 \text{ L}}{1000 \text{ mL}} = 0.1526 \text{ L}$$

$$M = \frac{0.01728 \text{ mol Na}^+}{0.1526 \text{ L}} = 0.113 \text{ } M \text{ Na}^+ \text{ ion}$$

c.   molar masses: NaCl, 58.44; KCl, 74.55 g; $CaCl_2$, 111.0 g

The total chloride ion in the solution comes from three different sources: we need to determine the number of moles of chloride ion contributed to the solution from each of the sources.

$$\text{from NaCl: mol Cl}^- = \text{mol NaCl} = 1.01 \text{ g} \times \frac{1 \text{ mol}}{58.44 \text{ g}} = 0.01728 \text{ mol}$$

$$\text{from KCl: mol Cl}^- = \text{mol KCl} = 2.11 \text{ g} \times \frac{1 \text{ mol}}{74.55 \text{ g}} = 0.02830 \text{ mol}$$

from $CaCl_2$: mol $Cl^- = 2 \times$ mol $CaCl_2 = 2 \times 1.55$ g $\times \dfrac{1 \text{ mol}}{111.0 \text{ g}}$

$$= 0.02792 \text{ mol}$$

total mol $Cl^- = 0.01728 + 0.02830 + 0.02792 = 0.0735$ mol

volume of solution (from part b) = 0.1526 L

$$M = \dfrac{0.0735 \text{ mol } Cl^-}{0.1526 \text{ L}} = 0.482 \ M$$

36. a. $\dfrac{(0.501 \ M)(172 \text{ mL})}{1000 \text{ mL}} = 0.0862 \ M$

b. $\dfrac{(0.115 \ M)(385 \text{ mL})}{1000 \text{ mL}} = 0.0443 \ M$

c. $\dfrac{(0.751 \ M)(0.285 \text{ L})}{1.00 \text{ L}} = 0.214 \ M$

38. a. $\dfrac{(41.5 \text{ mL})(0.118 \ M)(1)}{(0.242 \ M)(2)} = 10.1$ mL H_2SO_4

b. $\dfrac{(27.1 \text{ mL})(0.121 \ M)(3)}{(0.242 \ M)(2)} = 20.3$ mL H_2SO_4

Chapter 15   Acids and Bases

2.   In the Arrhenius definition, an acid is a substance that produces hydrogen ions (H^+) when dissolved in water, whereas a base is a substance that produces hydroxide ions (OH^-) in aqueous solution. These definitions proved to be too restrictive because the only base permitted was hydroxide ion, and the only solvent permitted was water.

4.   Conjugate acid-base pairs differ from each other by one proton (one hydrogen ion, H^+). For example, CH_3COOH (acetic acid), differs from its conjugate base, CH_3COO^- (acetate ion), by a single H^+ ion.

$$CH_3COOH\,(aq) \rightarrow CH_3COO^-\,(aq) + H^+\,(aq)$$

6.   When an acid is dissolved in water, the hydronium ion (H_3O^+) is formed. The hydronium ion is the conjugate *acid* of water (H_2O).

8.   a. HSO_4^- and SO_4^{2-} represent a conjugate acid-base pair (HSO_4^- is the acid, SO_4^{2-} is the base; they differ by one proton).

     b.   HBr and Br^- represent a conjugate acid-base pair (HBr is the acid, Br^- is the base; they differ by one proton.

     c. $H_2PO_4^-$ and PO_4^{3-} are not a conjugate acid-base pair; they differ by *two* protons ($H_2PO_4^-$ is the conjugate acid of HPO_4^{2-} and also the conjugate base of H_3PO_4; HPO_4^{2-} is the conjugate acid of PO_4^{3-}).

     d. HNO_3 and NO_2^- are not a conjugate acid-base pair; they differ by an oxygen atom as well as a proton (NO_3^- is the conjugate base of HNO_3; NO_2^- is the conjugate base of HNO_2).

10. a. NH_4^+ (acid), H_2O (base); NH_3 (base), H_3O^+ (acid)

     b. $HC_2H_3O_2$ (acid), H_2O (base); $C_2H_3O_2^-$ (base), H_3O^+ (acid)

     c. CH_3NH_2 (base), H_2O (acid); $CH_3NH_3^+$ (acid), OH^- (base)

12. The conjugate *acid* of the species indicated would have *one additional proton*:

     a. H_2CO_3

     b.   HBr

     c. $HClO_2$

     d. CH_3NH_2

14. The conjugate *bases* of the species indicated would have *one less proton*:

     a. CO_3^{2-}

     b. BrO_4^-

     c. SO_3^{2-}

     d. $C_2H_3O_2^-$

16.   a. $HBrO_3 + H_2O \rightarrow BrO_3^- + H_3O^+$

 b. $HI + H_2O \rightarrow I^- + H_3O^+$

 c. $HNO_3 + H_2O \rightarrow NO_3^- + H_3O^+$

 d. $NH_4^+ + H_2O \rightleftarrows NH_3 + H_3O^+$

18.   To say that an acid is *weak* in aqueous solution means that the acid does not easily transfer protons to water (and does not fully ionize). If an acid does not lose protons easily, then the acid's anion must be a strong attractor of protons (good at holding on to protons).

20.   A strong acid is one that loses its protons easily and fully ionizes in water; this means that the acid's conjugate base must be poor at attracting and holding on to protons, and is therefore a relatively weak base. A weak acid is one that resists loss of its protons and does not ionize well in water; this means that the acid's conjugate base attracts and holds onto protons tightly and is a relatively strong base.

22. H_2SO_4 (sulfuric): $H_2SO_4 + H_2O \rightarrow HSO_4^- + H_3O^+$

 HCl (hydrochloric): $HCl + H_2O \rightarrow Cl^- + H_3O^+$

 HNO_3 (nitric): $HNO_3 + H_2O \rightarrow NO_3^- + H_3O^+$

 $HClO_4$ (perchloric): $HClO_4 + H_2O \rightarrow ClO_4^- + H_3O^+$

24.   oxyacids: $HClO_4$, HNO_3, H_2SO_4, CH_3COOH, etc.

 non-oxyacids: HCl, HBr, HF, HI, HCN, etc.

26.   Bases that are *weak* have relatively strong conjugate acids:

 a. SO_4^{2-} is a moderately weak base; HSO_4^- is a moderately strong acid.

 b. Br^- is a very weak base; HBr is a strong acid.

 c. CN^- is a fairly strong base; HCN is a weak acid.

 d. CH_3COO^- is a fairly strong base; CH_3COOH is a weak acid.

28.   For example, HCO_3^- can behave as an acid if it reacts with something that more strongly gains protons than does HCO_3^- itself. For example, HCO_3^- would behave as an acid when reacting with hydroxide ion (a much stronger base).

$$HCO_3^- (aq) + OH^- (aq) \rightarrow CO_3^{2-} (aq) + H_2O(l).$$

On the other hand, HCO_3^- would behave as a base when reacted with something that more readily loses protons than does HCO_3^- itself. For example, HCO_3^- would behave as a base when reacting with hydrochloric acid (a much stronger acid).

$$HCO_3^- (aq) + HCl(aq) \rightarrow H_2CO_3(aq) + Cl^-(aq)$$

For $H_2PO_4^-$, similar equations can be written:

$$H_2PO_4^-(aq) + OH^-(aq) \rightarrow HPO_4^{2-}(aq) + H_2O(l)$$

$$H_2PO_4^-(aq) + H_3O^+(aq) \rightarrow H_3PO_4(aq) + H_2O(l)$$

30. The hydrogen ion concentration and the hydroxide ion concentration of water are *not* independent: they are related by the equilibrium

$$H_2O(l) \rightleftarrows H^+(aq) + OH^-(aq)$$

for which $K_w = [H^+][OH^-] = 1.0 \times 10^{-14}$ at 25°C.

If the concentration of one of these ions is increased by addition of a reagent producing H^+ or OH^-, then the concentration of the complementary ion will have to decrease so that the value of K_w will hold true. So if an acid is added to a solution, the concentration of hydroxide ion in the solution will decrease to a lower value. Similarly, if a base is added to a solution, then the concentration of hydrogen ion will have to decrease to a lower value.

32. $$K_w = [H^+][OH^-] = 1.0 \times 10^{-14} \text{ at } 25°C$$

$$[H^+] = \frac{1.0 \times 10^{-14}}{[OH^-]}$$

a. $$[H^+] = \frac{1.0 \times 10^{-14}}{3.99 \times 10^{-5} \, M} = 2.5 \times 10^{-10} \, M; \text{ solution is basic}$$

b. $$[H^+] = \frac{1.0 \times 10^{-14}}{2.91 \times 10^{-9} \, M} = 3.4 \times 10^{-6} \, M; \text{ solution is acidic}$$

c. $$[H^+] = \frac{1.0 \times 10^{-14}}{7.23 \times 10^{-2} \, M} = 1.4 \times 10^{-13} \, M; \text{ solution is basic}$$

d. $$[H^+] = \frac{1.0 \times 10^{-14}}{9.11 \times 10^{-7} \, M} = 1.1 \times 10^{-8} \, M; \text{ solution is basic}$$

34. $$[OH^-] = \frac{1.0 \times 10^{-14}}{[H^+]}$$

a. $$[OH^-] = \frac{1.0 \times 10^{-14}}{1.34 \times 10^{-2} \, M} = 7.5 \times 10^{-13} \, M; \text{ solution is acidic}$$

b. $$[OH^-] = \frac{1.0 \times 10^{-14}}{6.99 \times 10^{-7} \, M} = 1.4 \times 10^{-8} \, M; \text{ solution is acidic}$$

c. $$[OH^-] = \frac{1.0 \times 10^{-14}}{4.01 \times 10^{-9} \, M} = 2.5 \times 10^{-6} \, M; \text{ solution is basic}$$

d. $[OH^-] = \dfrac{1.0 \times 10^{-14}}{4.02 \times 10^{-13} \ M} = 2.5 \times 10^{-2} \ M$; solution is basic

36.   a. $[H^+] = 1.04 \times 10^{-8} \ M$ is more basic

b. $[OH^-] = 4.49 \times 10^{-6} \ M$ is more basic

c. $[OH^-] = 6.01 \times 10^{-7} \ M$ is more basic

38.   household ammonia (pH 12); blood (pH 7-8); milk (pH 6-7); vinegar (pH 3); lemon juice (pH 2-3); stomach acid (pH 2)

40.   The pH of a solution is defined as the *negative* of the logarithm of the hydrogen ion concentration, $pH = -\log[H^+]$. Mathematically, the *negative sign* in the definition causes the pH to *decrease* as the hydrogen ion concentration *increases*.

42. $pH = -\log[H^+]$

a. $pH = -\log[0.0010 \ M] = 3.000$; solution is acidic

b. $pH = -\log[2.19 \times 10^{-4} \ M] = 3.660$; solution is acidic

c. $pH = -\log[9.18 \times 10^{-11} \ M] = 10.037$; solution is basic

d. $pH = -\log[4.71 \times 10^{-7} \ M] = 6.327$; solution is acidic

44. $pH = -\log[H_3O^+]$ $pOH = -\log[OH^-]$ $pH = 14 - pOH$

a. $pH = -\log[3.99 \times 10^{-6} \ M] = 5.399$; solution is acidic

b. $pOH = -\log[4.21 \times 10^{-8} \ M] = 7.376$

$pH = 14 - 7.376 = 6.624$; solution is acidic

c. $pH = -\log[8.25 \times 10^{-11} \ M] = 10.084$; solution is basic

d. $pOH = -\log[9.21 \times 10^{-3} \ M] = 2.036$

$pH = 14 - 2.036 = 11.964$; solution is basic

46. $pOH = 14 - pH$

a. $pOH = 14 - 7.45 = 6.55$; solution is basic

b. $pOH = 14 - 1.89 = 12.11$; solution is acidic

c. $pOH = 14 - 13.15 = 0.85$; solution is basic

d. $pOH = 14 - 5.55 = 8.45$; solution is acidic

48. a.   pOH = $-\log[4.01 \times 10^{-2}\ M]$ = 1.397; solution is basic

$$[H_3O^+] = \frac{1.0 \times 10^{-14}}{4.01 \times 10^{-2}\ M} = 2.5 \times 10^{-13}\ M$$

b.   pOH = $-\log[9.87 \times 10^{-9}\ M]$ = 8.006; solution is acidic

$$[H_3O^+] = \frac{1.0 \times 10^{-14}}{9.87 \times 10^{-9}\ M} = 1.0 \times 10^{-6}\ M$$

c.   pOH = $-\log[5.23 \times 10^{-5}\ M]$ = 4.281; solution is basic

$$[H_3O^+] = \frac{1.0 \times 10^{-14}}{5.23 \times 10^{-5}\ M} = 1.9 \times 10^{-10}\ M$$

d.   pOH = $-\log[6.29 \times 10^{-12}\ M]$ = 11.201; solution is acidic

$$[H_3O^+] = \frac{1.0 \times 10^{-14}}{6.29 \times 10^{-12}\ M} = 1.6 \times 10^{-3}\ M$$

50. $[H^+]$ = {inv}{log}[-pH]

a. $[H^+]$ = {inv}{log}[-1.04] = $9.2 \times 10^{-2}\ M$

b. $[H^+]$ = {inv}{log}[-13.1] = $8 \times 10^{-14}\ M$

c. $[H^+]$ = {inv}{log}[-5.99] = $1.0 \times 10^{-6}\ M$

d. $[H^+]$ = {inv}{log}[-8.62] = $2.4 \times 10^{-9}\ M$

52. a. $[OH^-]$ = {inv}{log}[-7.00] = $1.0 \times 10^{-7}\ M$

b. $[OH^-]$ = {inv}{log}[-12.91] = $1.2 \times 10^{-13}\ M$

c. $[OH^-]$ = {inv}{log}[-1.82] = $1.5 \times 10^{-2}\ M$

d. $[OH^-]$ = {inv}{log}[-9.41] = $3.9 \times 10^{-10}\ M$

54. a.   pH = 14 - pOH = 14 - 11.31 = 2.69

b.   pOH = $-\log[7.22 \times 10^{-5}\ M]$ = 4.141

pH = 14 - pOH = 14 - 4.141 = 9.859 (9.86)

c.   pH = $-\log[9.93 \times 10^{-4}\ M]$ = 3.003

d.   pOH = $-\log[1.49 \times 10^{-8}\ M]$ = 7.827

pH = 14 - pOH = 14 - 7.827 = 6.173 (6.17)

56.   The solution contains water molecules, H_3O^+ ions (protons), and NO_3^- ions. Because HNO_3 is a strong acid, which is completely ionized in water, there are no HNO_3 molecules present.

58. a. HNO_3 is a strong acid and completely ionized so

$[H^+] = 1.21 \times 10^{-3}$ M and pH = 2.917.

b. $HClO_4$ is a strong acid and completely ionized so

$[H^+] = 0.000199$ M and pH = 3.701.

c. HCl is a strong acid and completely ionized so

$[H^+] = 5.01 \times 10^{-5}$ M and pH = 4.300.

d. HBr is a strong acid and completely ionized so

$[H^+] = 0.00104$ M and pH = 2.983.

60. A buffered solution consists of a mixture of a weak acid and its conjugate base; one example of a buffered solution is a mixture of acetic acid (CH_3COOH) and sodium acetate ($NaCH_3COO$).

62. The weak acid component of a buffered solution is capable of reacting with added strong base. For example, using the buffered solution given as an example in Question 60, acetic acid would consume added sodium hydroxide as follows:

$$CH_3COOH(aq) + NaOH(aq) \rightarrow NaCH_3COO(aq) + H_2O(l).$$

Acetic acid *neutralizes* the added NaOH and prevents it from having much effect on the overall pH of the solution.

64. added NaOH: $CH_3COOH + OH^- \rightarrow CH_3COO^- + H_2O$

added HCl: $CH_3COO^- + H_3O^+ \rightarrow CH_3COOH + H_2O$

added NaOH: $H_2S + OH^- \rightarrow HS^- + H_2O$

added HCl: $HS^- + H_3O^+ \rightarrow H_2S + H_2O$

66. a. NaOH is completely ionized, so $[OH^-] = 0.10$ M.

pOH = $-\log[0.10]$ = 1.00

pH = 14 − 1.00 = 13.00

b. KOH is completely ionized, so $[OH^-] = 2.0 \times 10^{-4}$ M.

pOH = $-\log[2.0 \times 10^{-4}]$ = 3.70

pH = 14 − 3.70 = 10.30

c. CsOH is completely ionized, so $[OH^-] = 6.2 \times 10^{-3}$ M.

pOH = $-\log[6.2 \times 10^{-3}]$ = 2.21

pH = 14 − 2.21 = 11.79

d.   NaOH is completely ionized, so $[OH^-] = 0.0001$ *M*.

pOH = $-\log[0.0001]$ = 4.0

pH = 14 - 4.0 = 10.0

68.   *a*, *b*, and *d* represent basic solutions; *c* represents an *acidic* solution because there is less hydroxide ion than hydrogen ion.

70.   *a*, *c*, and *e* represent strong acids; *b* and *d* are typical *weak* acids.

72.   Ordinarily in calculating the pH of strong acid solutions, the major contribution to the concentration of hydrogen ion present is from the dissolved strong acid; we ordinarily neglect the small amount of hydrogen ion present in such solutions due to the ionization of water. With 1.0×10^{-7} *M* HCl solution, however, the amount of hydrogen ion present due to the ionization of *water* is *comparable* to that present due to the addition of *acid* (HCl) and must be considered in the calculation of pH.

74.   accepts

76.   base

78.   carboxyl (-COOH)

80. 1.0×10^{-14}

82.   higher

84.   pH

86.   weak acid

88.   a. H_2O and OH^- represent a conjugate acid-base pair (H_2O is the acid, having one more proton than the base, OH^-).

b. H_2SO_4 and SO_4^{2-} are *not* a conjugate acid-base pair (they differ by *two* protons). The conjugate base of H_2SO_4 is HSO_4^-; the conjugate acid of SO_4^{2-} is also HSO_4^-.

c. H_3PO_4 and $H_2PO_4^-$ represent a conjugate acid-base pair (H_3PO_4 is the acid, having one more proton than the base $H_2PO_4^-$).

d. $HC_2H_3O_2$ and $C_2H_3O_2^-$ represent a conjugate acid-base pair ($HC_2H_3O_2$ is the acid, having one more proton than the base $C_2H_3O_2^-$).

90.   The conjugate *acid* of the species indicated would have *one additional proton*:

   a. NH_4^+

   b. NH_3

   c. H_3O^+

   d. H_2O

92.   When an acid ionizes in water, a proton is released to the water as an H_3O^+ ion:

   a. $CH_3CH_2COOH + H_2O \rightleftharpoons CH_3CH_2COO^- + H_3O^+$

   b. $NH_4^+ + H_2O \rightleftharpoons NH_3 + H_3O^+$

   c. $H_2SO_4 + H_2O \rightarrow HSO_4^- + H_3O^+$

   d. $H_3PO_4 + H_2O \rightleftharpoons H_2PO_4^- + H_3O^+$

94. $K_w = [H^+][OH^-] = 1.0 \times 10^{-14}$ at 25°C

   $$[H^+] = \frac{1.0 \times 10^{-14}}{[OH^-]}$$

   a. $[H^+] = \dfrac{1.0 \times 10^{-14}}{4.22 \times 10^{-3} \ M} = 2.4 \times 10^{-12} \ M$; solution is basic

   b. $[H^+] = \dfrac{1.0 \times 10^{-14}}{1.01 \times 10^{-13} \ M} = 9.9 \times 10^{-2} \ M$; solution is acidic

   c. $[H^+] = \dfrac{1.0 \times 10^{-14}}{3.05 \times 10^{-7} \ M} = 3.3 \times 10^{-8} \ M$; solution is basic

   d. $[H^+] = \dfrac{1.0 \times 10^{-14}}{6.02 \times 10^{-6} \ M} = 1.7 \times 10^{-9} \ M$; solution is basic

96.   a. $[OH^-] = 0.0000032 \ M$ is more basic

   b. $[OH^-] = 1.54 \times 10^{-8} \ M$ is more basic

   c. $[OH^-] = 4.02 \times 10^{-7} \ M$ is more basic

98. $pOH = -\log[OH^-]$ $pH = 14 - pOH$

   a. $pOH = -\log[1.4 \times 10^{-6} \ M] = 5.85$

   $pH = 14 - 5.85 = 8.15$; solution is basic

   b. $pOH = -\log[9.35 \times 10^{-9} \ M] = 8.029 = 8.03$

   $pH = 14 - 8.029 = 5.971 = 5.97$; solution is acidic

c. $pOH = -\log[2.21 \times 10^{-1} \ M] = 0.656 = 0.66$

$pH = 14 - 0.656 = 13.344 = 13.34$; solution is basic

d. $pOH = -\log[7.98 \times 10^{-12} \ M] = 11.098 = 11.10$

$pH = 14 - 11.098 = 2.902 = 2.90$; solution is acidic

100. a. $[OH^-] = \dfrac{1.0 \times 10^{-14}}{5.72 \times 10^{-4} \ M} = 1.75 \times 10^{-11} \ M = 1.8 \times 10^{-11} \ M$

$pOH = -\log[1.75 \times 10^{-11} \ M] = 10.76$

$pH = 14 - 10.76 = 3.24$

b. $[H^+] = \dfrac{1.0 \times 10^{-14}}{8.91 \times 10^{-5} \ M} = 1.12 \times 10^{-10} \ M = 1.1 \times 10^{-10} \ M$

$pH = -\log[1.12 \times 10^{-10} \ M] = 9.95$

$pOH = 14 - 9.95 = 4.05$

c. $[OH^-] = \dfrac{1.0 \times 10^{-14}}{2.87 \times 10^{-12} \ M} = 3.48 \times 10^{-3} \ M = 3.5 \times 10^{-3} \ M$

$pOH = -\log[3.48 \times 10^{-3} \ M] = 2.46$

$pH = 14 - 2.46 = 11.54$

d. $[H^+] = \dfrac{1.0 \times 10^{-14}}{7.22 \times 10^{-8} \ M} = 1.39 \times 10^{-7} \ M = 1.4 \times 10^{-7} \ M$

$pH = -\log[1.39 \times 10^{-7} \ M] = 6.86$

$pOH = 14 - 6.86 = 7.14$

102. $pH = 14 - pOH$ $[H^+] = \{inv\}\{log\}[-pH]$

a. $[H^+] = \{inv\}\{log\}[-5.41] = 3.9 \times 10^{-6} \ M$

b. $pH = 14 - 12.04 = 1.96$

$[H^+] = \{inv\}\{log\}[-1.96] = 1.1 \times 10^{-2} \ M$

c. $[H^+] = \{inv\}\{log\}[-11.91] = 1.2 \times 10^{-12} \ M$

d. $pH = 14 - 3.89 = 10.11$

$[H^+] = \{inv\}\{log\}[-10.11] = 7.8 \times 10^{-11} \ M$

104. a. $HClO_4$ is a strong acid and completely ionized so

$[H^+] = 1.4 \times 10^{-3} \ M$ and $pH = 2.85.$

b.   HCl is a strong acid and completely ionized so

$[H^+]$ = 3.0×10^{-5} M and pH = 4.52.

c. HNO_3 is a strong acid and completely ionized so

$[H^+]$ = 5.0×10^{-2} M and pH = 1.30.

d.   HCl is a strong acid and completely ionized so

$[H^+]$ = 0.0010 M and pH = 3.00.

Chapter 16   Equilibrium

2.   The carbon-oxygen bonds in two carbon monoxide molecules and the
      oxygen-oxygen bond in an oxygen gas molecule must break, and the carbon-
      oxygen bonds in two carbon dioxide molecules must form.

4.   The symbol E_a represents the *activation energy* of the reaction. The
      activation energy is the minimum energy two colliding molecules must
      possess in order for the collision to result in reaction. If molecules
      do not possess energies equal to or greater than E_a, a collision between
      these molecules will not result in a reaction.

6.   Enzymes are biochemical catalysts that accelerate the complicated
      biochemical reactions in cells that would ordinarily be too slow to
      sustain life at normal body temperatures.

8.   A state of equilibrium is attained when two opposing processes are
      exactly balanced so there is no further observable net change in the
      system.

10.   Chemical equilibrium occurs when two *opposing* chemical reactions reach
      the *same speed* in a closed system. When a state of chemical equilibrium
      has been reached, the concentrations of reactants and products present
      in the system remain *constant* with time, and the reaction appears to
      "stop." A chemical reaction that reaches a state of equilibrium is
      indicated by using a double arrow ($\rightleftharpoons$). The points of the double arrow
      point in opposite directions, to indicate that two opposite processes
      are going on.

12.   Although we recognize a state of chemical equilibrium by the fact that
      the concentrations of reactants and products no longer change with time,
      the lack of change results from the fact that two *opposing* processes are
      going on at the same time and rate (not because the reaction has truly
      "stopped"). Further reaction in the forward direction is canceled out by
      an equal extent of reaction in the reverse direction. This dynamic model
      of equilibrium has been tested using radioactive tracer atoms.

14.   The equilibrium constant is a *ratio* of concentration of products to
      concentration of reactants, with all concentrations measured at
      equilibrium. Depending on the amount of reactant present at the
      beginning of an experiment, there may be different absolute amounts of
      reactants and products present at equilibrium, but the *ratio* will always
      be the same for a given reaction at a given temperature. For example,
      the ratios (4/2) and (6/3) are different absolutely in terms of the
      numbers involved, but each of these ratios has the *value* of 2.

16.   a. $K = \dfrac{[NO_2]\,[O_2]}{[NO]\,[O_3]}$

      b. $K = \dfrac{[SO_3]\,[NO]}{[SO_2]\,[NO_2]}$

      c. $K = \dfrac{[HCl]^4\,[O_2]}{[Cl_2]^2\,[H_2O]^2}$

18. a. $K = \dfrac{[CH_3OH]}{[CO][H_2]^2}$

    b. $K = \dfrac{[NO]^2[O_2]}{[NO_2]^2}$

    c. $K = \dfrac{[PBr_3]^4}{[P_4][Br_2]^6}$

20. $NO(g) + O_3(g) \rightleftharpoons NO_2(g) + O_2(g)$

    $K = \dfrac{[NO_2][O_2]}{[NO][O_3]} = \dfrac{[3.5 \times 10^{-2}\ M][7.8 \times 10^{-3}\ M]}{[2.1 \times 10^{-3}\ M][4.7 \times 10^{-5}\ M]} = 2.8 \times 10^3$

22. $2NO(g) + 2H_2(g) \rightleftharpoons N_2(g) + 2H_2O(g)$

    $K = \dfrac{[N_2][H_2O]^2}{[NO]^2[H_2]^2} = \dfrac{[5.3 \times 10^{-2}\ M][2.9 \times 10^{-3}\ M]}{[8.1 \times 10^{-3}\ M]^2[4.1 \times 10^{-5}\ M]^2} = 4.0 \times 10^6$

24. Equilibrium constants represent ratios of the *concentrations* of products and reactants present at the point of equilibrium. The *concentration* of a pure solid or of a pure liquid is constant and is determined by the density of the solid or liquid. For example, suppose you had a liter of water. Within that liter of water are 55.5 mol of water (the number of moles of water that is contained in one liter of water *does not vary*).

26. a. $K = \dfrac{1}{[H_2]^2[CO]}$

    b. $K = \dfrac{1}{[H_2]}$

    c. $K = \dfrac{1}{[CO_2]}$

28. a. $K = [Br_2]^3[N_2]$

    b. $K = \dfrac{[H_2O]}{[H_2]}$

    c. $K = [CO_2]$

30. When an additional amount of one of the reactants is added to an equilibrium system, the system shifts to the right and adjusts so as to consume some of the added reactant. This results in a net *increase* in the amount of product, compared to the equilibrium system before the additional reactant was added, and so the amount of $CO_2(g)$ in the system will be higher than if the additional $CO(g)$ had not been added. The numerical *value* of the equilibrium constant does *not* change when a reactant is added: the concentrations of all reactants and products adjust until the correct value of K is once again achieved.

32.   If heat is applied to an endothermic reaction (i.e., the temperature is raised), the equilibrium is shifted to the right. More product will be present at equilibrium than if the temperature had not been increased. The value of K increases.

34.   a.   shifts right (system reacts to get rid of excess SO_2)

b.   shifts right (system reacts to replace missing SO_3)

c.   no change (catalysts do not affect the position of equilibrium)

36.   a.   no effect (UO_2 is a solid)

b.   no effect (Xe is not involved in the reaction)

c.   shifts to left (if HF attacks glass, it is removed from system, causing reaction to replace the lost HF)

d.   shifts to right

e.   shifts to left (4 mol gas versus 3 mol gas)

38.   As the reaction is exothermic, heat is effectively a product of the reaction as written. Lowering the temperature of the system (removing heat from the system) tends to favor the formation of products, and shifts the equilibrium to the right.

40.   For an *endo*thermic reaction, an increase in temperature will shift the position of equilibrium to the right (toward products).

42.   Adding addition N_2 or H_2 would tend to favor products. Compressing the system to smaller volume would tend to favor products. The reaction is *exo*thermic and should be performed at as low a temperature as possible (consistent with the molecules still having sufficient energy to react).

44.   A large equilibrium constant means that the concentration of products is large, compared to the concentration of reactants. The position of equilibrium lies far to the right in favor of the products.

46. $$K = \frac{[NCl_3]^2}{[N_2][Cl_2]^3} = \frac{[1.9 \times 10^{-1}]^2}{[1.4 \times 10^{-3} \ M][4.3 \times 10^{-4} \ M]^3} = 3.2 \times 10^{11}$$

48. $K = [CO_2] = [2.1 \times 10^{-3} \ M] = 2.1 \times 10^{-3}$

50. $$K = \frac{[H_2]^2[O_2]}{[H_2O]^2}$$

$$2.4 \times 10^{-3} = \frac{[1.9 \times 10^{-2}]^2[O_2]}{[1.1 \times 10^{-1}]^2}$$

$$[O_2] = \frac{(2.4 \times 10^{-3})(1.1 \times 10^{-1})^2}{(1.9 \times 10^{-2})^2} = 8.0 \times 10^{-2} \ M$$

52. $K = \dfrac{[NO_2]^2}{[N_2O_4]} = 8.1 \times 10^{-3}$

$8.1 \times 10^{-3} = \dfrac{[0.0021 \ M]^2}{[N_2O_4]}$

$[N_2O_4] = 5.4 \times 10^{-4} \ M$

54.   solubility product, K_{sp}

56.   Stirring or grinding the solute increases the speed with which the solute dissolves, but the ultimate *amount* of solute that dissolves is fixed by the equilibrium constant for the dissolving process, K_{sp}, which changes only with temperature. Therefore only the temperature will affect the solubility.

58.   a. $PbBr_2(s) \ \rightleftarrows \ Pb^{2+}(aq) + 2Br^-(aq)$

$K_{sp} = [Pb^{2+}][Br^-]^2$

b. $Ag_2S(s) \ \rightleftarrows \ 2Ag^+(aq) + S^{2-}(aq)$

$K_{sp} = [Ag^+]^2[S^{2-}]$

c. $PbCO_3(s) \ \rightleftarrows \ Pb^{2+}(aq) + CO_3^{2-}(aq)$

$K_{sp} = [Pb^{2+}][CO_3^{2-}]$

d. $Sr_3(PO_4)_2(s) \ \rightleftarrows \ 3Sr^{2+}(aq) + 2PO_4^{3-}(aq)$

$K_{sp} = [Sr^{2+}]^3[PO_4^{3-}]^2$

60. $ZnCO_3(s) \ \rightleftarrows \ Zn^{2+}(aq) + CO_3^{2-}(aq)$

$K_{sp} = [Zn^{2+}][CO_3^{2-}]$

If 1.7×10^{-5} mol of $ZnCO_3$ dissolves per liter, then the concentrations of the two ions produced will each also be $1.7 \times 10^{-5} \ M$.

$K_{sp} = [1.7 \times 10^{-5} \ M][1.7 \times 10^{-5} \ M] = 2.9 \times 10^{-10}$.

62. $CuCrO_4(s) \ \rightleftarrows \ Cu^{2+}(aq) + CrO_4^{2-}(aq)$

$K_{sp} = [Cu^{2+}][CrO_4^{2-}]$

molar mass of $CuCrO_4 = 179.55$ g

1.1×10^{-5} g $CuCrO_4 \times \dfrac{1 \ mol}{179.55 \ g} = 6.1 \times 10^{-8}$ mol

If 6.1×10^{-8} mol of $CuCrO_4$ dissolve per liter, then the concentrations of $Cu^{2+}(aq)$ and $CrO_4^{2-}(aq)$ are also each 6.1×10^{-8} M.

$$K_{sp} = (6.1 \times 10^{-8} \, M)(6.1 \times 10^{-8} \, M) = 3.8 \times 10^{-15}$$

64. $BaCO_3(s) \rightleftarrows Ba^{2+}(aq) + CO_3^{2-}(aq)$

$$K_{sp} = [Ba^{2+}][CO_3^{2-}] = 8.2 \times 10^{-9}$$

Let x represent the number of moles of $BaCO_3(s)$ that dissolve per liter, then $[Ba^{2+}] = x$ and $[CO_3^{2-}] = x$ also from the stoichiometry of the reaction;

$$K_{sp} = [x][x] = x^2 = 8.2 \times 10^{-9}$$

$$x = [BaCO_3] = 9.1 \times 10^{-5} \, M$$

66. $HgS(s) \rightleftarrows Hg^{2+}(aq) + S^{2-}(aq)$

$$K_{sp} = [Hg^{2+}][S^{2-}] = 3 \times 10^{-53}$$

Let x represent the number of moles of $HgS(s)$ that dissolve per liter. Then $[Hg^{2+}] = x$ and $[S^{2-}] = x$ also from the stoichiometry of the reaction;

$$K_{sp} = [x][x] = x^2 = 3 \times 10^{-53}$$

$$x = [HgS] = 5 \times 10^{-27} \, M \text{ in the saturated solution.}$$

molar mass of $PbCrO_4$ = 323.2 g

5.3×10^{-7} mol/L $\times$ 323.2 g/mol = 1.7×10^{-4} g/L

68. $PbCl_2(s) \rightleftarrows Pb^{2+}(aq) + 2Cl^-(aq)$

$$K_{sp} = [Pb^{2+}][Cl^-]^2$$

If $PbCl_2$ dissolves to the extent of 3.6×10^{-2} M, then

$[Pb^{2+}] = 3.6 \times 10^{-2}$ M and

$[Cl^-] = 2 \times (3.6 \times 10^{-2}) = 7.2 \times 10^{-2}$ M.

$$K_{sp} = (3.6 \times 10^{-2} \, M)(7.2 \times 10^{-2} \, M)^2 = 1.9 \times 10^{-4}$$

molar mass $PbCl_2$ = 278.1 g

$$3.6 \times 10^{-2} \, \frac{mol}{1 \, L} \times \frac{278.1 \, g}{1 \, mol} = 10. \text{ g/L}$$

70. $Fe(OH)_3(s) \rightleftarrows Fe^{3+}(aq) + 3OH^-(aq)$

    $K_{sp} = [Fe^{3+}][OH^-]^3 = 4 \times 10^{-38}$

    Let x represent the number of moles of $Fe(OH)_3$ that dissolve per liter; then $[Fe^{3+}] = x$.

    The amount of hydroxide ion that would be produced by the dissolving of the $Fe(OH)_3$ would then be $3x$, but pure water itself contains hydroxide ion at the concentration of 1.0×10^{-7} M (see Chapter 17). The total concentration of hydroxide ion is then $[OH^-] = (3x + 1.0 \times 10^{-7})$. As x must be a very small number [because $Fe(OH)_3$ is not very soluble], we can save ourselves a lot of arithmetic if we use the approximation that

    $(3x + 1.0 \times 10^{-7} M) = 1.0 \times 10^{-7}$

    $K_{sp} = [x][1.0 \times 10^{-7}]^3 = 4 \times 10^{-38}$

    $x = 4 \times 10^{-17} M$

    molar mass $Fe(OH)_3 = 106.9$ g

    $$\frac{4 \times 10^{-17} \text{ mol}}{1.00 \text{ L}} \times \frac{106.9 \text{ g}}{1 \text{ mol}} = 4 \times 10^{-15} \text{ g/L}$$

72. An increase in temperature increases the fraction of molecules that possess sufficient energy for a collision to result in a reaction.

74. catalyst

76. constant

78. When we say that a chemical equilibrium is *dynamic*, we are recognizing the fact that even though the reaction has appeared macroscopically to have stopped, on a microscopic basis the forward and reverse reactions are still taking place, at the same speed.

80. heterogeneous

82. position

84. Heat is considered a *product* of an exothermic process. Adding a product to a system in equilibrium causes the reverse reaction to occur (producing additional reactants).

86. An equilibrium reaction may come to many *positions* of equilibrium, but at each possible position of equilibrium, the numerical value of the equilibrium constant is fulfilled. If different amounts of reactant are taken in different experiments, the *absolute amounts* of reactant and

product present at the point of equilibrium reached will differ from one experiment to another, but the *ratio* that defines the equilibrium constant will be the same.

88. $PCl_5(g) \rightleftarrows PCl_3(g) + Cl_2(g)$

$$K = \frac{[PCl_3][Cl_2]}{[PCl_5]} = 4.5 \times 10^{-3}$$

The concentration of PCl_5 is to be twice the concentration of PCl_3:

$$[PCl_5] = 2 \times [PCl_3]$$

$$K = \frac{[PCl_3][Cl_2]}{2 \times [PCl_3]} = 4.5 \times 10^{-3}$$

$$K = \frac{[Cl_2]}{2} = 4.5 \times 10^{-3} \quad \text{and} \quad [Cl_2] = 9.0 \times 10^{-3} \ M$$

90.   As all of the metal carbonates indicated have the metal ion in the +2 oxidation state, we can illustrate the calculations for a general metal carbonate, MCO_3:

$$MCO_3(s) \rightleftarrows M^{2+}(aq) + CO_3^{2-}(aq) \qquad K_{sp} = [M^{2+}(aq)][CO_3^{2-}(aq)]$$

If we then let x represent the number of moles of MCO_3 that dissolve per liter, then $[M^{2+}(aq)] = x$ and $[CO_3^{2-}(aq)] = x$ also because the reaction is of 1:1 stoichiometry. Therefore,

$K_{sp} = [M^{2+}(aq)][CO_3^{2-}(aq)] = x^2$ for each salt. Solving for x gives the following results.

$$[BaCO_3] = x = 7.1 \times 10^{-5} \ M$$
$$[CdCO_3] = x = 2.3 \times 10^{-6} \ M$$
$$[CaCO_3] = x = 5.3 \times 10^{-5} \ M$$
$$[CoCO_3] = x = 3.9 \times 10^{-7} \ M$$

92.   Although a small solubility product generally implies a small solubility, comparisons of solubility based directly on K_{sp} values are only valid if the salts produce the same numbers of positive and negative ions per formula when they dissolve. For example, one can compare the solubilities of $AgCl(s)$ and $NiS(s)$ directly using K_{sp}, because each salt produces one positive and one negative ion per formula when dissolved. One could not directly compare $AgCl(s)$ with a salt such as $Ca_3(PO_4)_2$, however.

94.   At higher temperatures, the average kinetic energy of the reactant molecules is larger. At higher temperatures, the probability that a collision between molecules will be energetic enough for reaction to take place is larger. On a molecular basis, a higher temperature means a given molecule will be moving faster.

96. a. $$K = \frac{[HBr]^2}{[H_2][Br_2]}$$

     b. $$K = \frac{[H_2S]^2}{[H_2]^2[S_2]}$$

     c. $$K = \frac{[HCN]^2}{[H_2][C_2N_2]}$$

98. $$K = \frac{[Br]^2}{[Br_2]} = \frac{(0.034 \; M)^2}{(0.97 \; M)} = 1.2 \times 10^{-3}$$

100. a. $$K = \frac{1}{[O_2]^3}$$

     b. $$K = \frac{1}{[NH_3][HCl]}$$

     c. $$K = \frac{1}{[O_2]}$$

102. An *exothermic* reaction is one that liberates heat energy. Increasing the temperature (adding heat) for such a reaction is fighting against the reaction's own tendency to liberate heat. The net effect of raising the temperature will be a shift to the left to decrease the amount of product. If it is desired to increase the amount of product in an exothermic reaction, heat must be *removed* from the system. Changing the temperature *does* change the numerical value of the equilibrium constant for a reaction.

104. The reaction is *exothermic* as written. An increase in temperature (addition of heat) will shift the reaction to the left (toward reactants).

106. $$K = \frac{[NH_3]^2}{[N_2][H_2]^3}$$

     $$1.3 \times 10^{-2} = \frac{[NH_3]^2}{(0.10 \; M)(0.10 \; M)^3}$$

     $$[NH_3]^2 = 1.3 \times 10^{-2} \times (0.10) \times (0.10)^3 = 1.3 \times 10^{-6}$$

     $$[NH_3] = 1.1 \times 10^{-3} \; M$$

108. a. $Cu(OH)_2(s) \rightleftharpoons Cu^{2+}(aq) + 2OH^-(aq)$
          $K_{sp} = [Cu^{2+}][OH^-]^2$

     b. $Cr(OH)_3(s) \rightleftharpoons Cr^{3+}(aq) + 3OH^-(aq)$
          $K_{sp} = [Cr^{3+}][OH^-]^3$

     c. $Ba(OH)_2(s) \rightleftharpoons Ba^{2+}(aq) + 2OH^-(aq)$
          $K_{sp} = [Ba^{2+}][OH^-]^2$

d. $Sn(OH)_2(s) \rightleftharpoons Sn^{2+}(aq) + 2OH^-(aq)$
      $K_{sp} = [Sn^{2+}][OH^-]^2$

110.   molar mass AgCl = 143.4 g

$$9.0 \times 10^{-4} \text{ g AgCl/L} \times \frac{1 \text{ mol AgCl}}{143.4 \text{ g AgCl}} = 6.28 \times 10^{-6} \text{ mol AgCl/L}$$

$AgCl(s) \rightleftharpoons Ag^+(aq) + Cl^-(aq)$

$K_{sp} = [Ag^+][Cl^-] = (6.28 \times 10^{-6} M)(6.28 \times 10^{-6} M) = 3.9 \times 10^{-11}$

112.   molar mass $Ni(OH)_2$ = 92.71 g

$$\frac{0.14 \text{ g Ni(OH)}_2}{1.00 \text{ L}} \times \frac{1 \text{ mol Ni(OH)}_2}{92.71 \text{ g Ni(OH)}_2} = 1.510 \times 10^{-3} M$$

$Ni(OH)_2(s) \rightleftharpoons Ni^{2+}(aq) + 2OH^-(aq)$

$K_{sp} = [Ni^{2+}][OH^-]^2$

If $1.510 \times 10^{-3} M$ of $Ni(OH)_2$ dissolves, then

$[Ni^{2+}] = 1.510 \times 10^{-3} M$ and

$[OH^-] = 2 \times (1.510 \times 10^{-3} M) = 3.020 \times 10^{-3} M.$

$K_{sp} = (1.510 \times 10^{-3} M)(3.020 \times 10^{-3} M)^2 = 1.4 \times 10^{-8}$

114.   The activation energy is the minimum energy two colliding molecules must possess in order for the collision to result in reaction. If molecules do not possess energies equal to or greater than E_a, a collision between these molecules will not result in a reaction.

116.   Once a system has reached equilibrium the net concentration of product no longer increases because molecules of product already present react to form the original reactants. This is not to say that the *same* product molecules are necessarily always present.

118.   a. $K = [H_2O(g)][CO_2(g)]$

b. $K = [CO_2]$

c. $K = \dfrac{1}{[O_2]^3}$

Cumulative Review: Chapters 15 and 16

2.   A conjugate acid-base pair consists of two species related to each other
by donation or acceptance of a single proton, H^+. An acid has one more
H^+ than its conjugate base; a base has one less H^+ than its conjugate
acid.

Brönsted-Lowry acids:

$$HCl(aq) + H_2O(l) \rightarrow Cl^-(aq) + H_3O^+(aq)$$

$$H_2SO_4(aq) + H_2O(l) \rightarrow HSO_4^-(aq) + H_3O^+(aq)$$

$$H_3PO_4(aq) + H_2O(l) \rightleftarrows H_2PO_4^-(aq) + H_3O^+(aq)$$

$$NH_4^+(aq) + H_2O(l) \rightleftarrows NH_3(aq) + H_3O^+(aq)$$

Brönsted-Lowry bases:

$$NH_3(aq) + H_2O(l) \rightleftarrows NH_4^+(aq) + OH^-(aq)$$

$$HCO_3^-(aq) + H_2O(l) \rightleftarrows H_2CO_3(aq) + OH^-(aq)$$

$$NH_2^-(aq) + H_2O(l) \rightarrow NH_3(aq) + OH^-(aq)$$

$$H_2PO_4^-(aq) + H_2O(l) \rightleftarrows H_3PO_4(aq) + OH^-(aq)$$

4.   The strength of an acid is a direct result of the position of the acid's
ionization equilibrium. Strong acids are those whose ionization
equilibrium positions lie far to the right, whereas weak acids are those
whose equilibrium positions lie only slightly to the right. For example,
HCl, HNO_3, and $HClO_4$ are all strong acids, which means they are
completely ionized in aqueous solution (the position of equilibrium is
very far to the right):

$$HCl(aq) + H_2O(l) \rightarrow Cl^-(aq) + H_3O^+(aq)$$

$$HNO_3(aq) + H_2O(l) \rightarrow NO_3^-(aq) + H_3O^+(aq)$$

$$HClO_4(aq) + H_2O(l) \rightarrow ClO_4^-(aq) + H_3O^+(aq)$$

As these are very strong acids, we know their anions (Cl^-, NO_3^-, ClO_4^-)
must be very weak bases, and that solutions of the sodium salts of these
anions would *not* be appreciably basic. As these acids have a strong
tendency to lose protons, there is very little tendency for the anions
(bases) to gain protons.

6.   The pH of a solution is defined as the negative of the base 10 logarithm
of the hydrogen ion concentration in the solution; that is

$$pH = -\log_{10}[H^+].$$

As in pure water, the amount of $H^+(aq)$ ion present is equal to the
amount of $OH^-(aq)$ ion, we say that pure water is *neutral*. As $[H^+] = 1.0
\times 10^{-7}$ M in pure water, this means that the pH of pure water is

$-\log[1.0 \times 10^{-7} \; M] = 7.00$. Solutions in which the hydrogen ion concentration is greater than $1.0 \times 10^{-7} \; M$ (pH < 7.00) are *acidic*; solutions in which the hydrogen ion concentration is less than $1.0 \times 10^{-7} \; M$ (pH > 7.00) are *basic*. The pH scale is logarithmic. When the pH changes by one unit, this corresponds to a change in the hydrogen ion concentration by a factor of *ten*.

In some instances, it may be more convenient to speak directly about the hydroxide ion concentration present in a solution, and so an analogous logarithmic expression is defined for the hydroxide ion concentration:

$$pOH = -\log_{10}[OH^-].$$

The concentrations of hydrogen ion and hydroxide ion in water (and in aqueous solutions) are *not* independent of one another, but rather are related by the dissociation equilibrium constant for water,

$$K_w = [H^+][OH^-] = 1.0 \times 10^{-14} \text{ at } 25°C.$$

From this constant it is obvious that pH + pOH = 14.00 for water (or an aqueous solution) at 25°C.

8.   Chemists envision that a reaction can only take place between molecules if the molecules physically *collide* with each other. Furthermore, when molecules collide, the molecules must collide with enough force for the reaction to be successful (there must be enough energy to break bonds in the reactants), and the colliding molecules must be positioned with the correct relative orientation for the products (or intermediates) to form. Reactions tend to be faster if higher concentrations are used for the reaction; because, if there are more molecules present per unit volume there will be more collisions between molecules in a given time period. Reactions are faster at higher temperatures because at higher temperatures the reactant molecules have a higher average kinetic energy, and the number of molecules that will collide with sufficient force to break bonds increases.

10.   Chemists define equilibrium as the exact balancing of two exactly opposing processes. When a chemical reaction is begun by combining pure reactants, the only process possible initially is

reactants → products

However, for many reactions, as the concentration of product molecules increases, it becomes more likely that product molecules will collide and react with each other

products → reactants

giving back molecules of the original reactants. At some point in the process the rates of the forward and reverse reactions become equal, and the system attains chemical equilibrium. To an outside observer, the system appears to have stopped reacting. On a microscopic basis, though, both the forward and reverse processes are still going on: every time additional molecules of the product form, however, somewhere else in the system molecules of product react to give back molecules of reactant.

Once the point is reached that product molecules are reacting at the same speed at which they are forming, there is no further net change in concentration. A graph showing how the rates of the forward and reverse reactions change with time is given in the text as Figure 16.8. At the start of the reaction, the rate of the forward reaction is at its maximum, whereas the rate of the reverse reaction is zero. As the reaction proceeds, the rate of the forward reaction gradually decreases as the concentration of reactants decreases, whereas the rate of the reverse reaction increases as the concentration of products increases. Once the two rates have become equal, the reaction has reached a state of equilibrium.

12. The equilibrium constant for a reaction is a *ratio* of the concentration of products present at the point of equilibrium to the concentration of reactants still present. A *ratio* means that we have one number divided by another number (for example, the density of a substance is the ratio of a substance's mass to its volume). As the equilibrium constant is a ratio, there are an infinite number of sets of data that can give the same ratio: for example, the ratios 8/4, 6/3, 100/50 all have the same value, 2. The actual concentrations of products and reactants will differ from one experiment to another involving a particular chemical reaction, but the ratio of the amount of product to reactant at equilibrium should be the same for each experiment.

Consider this simple example: suppose we have a reaction for which $K = 4$, and we begin this reaction with 100 reactant molecules. At the point of equilibrium, there should be 80 molecules of product and 20 molecules of reactant remaining (80/20 = 4). Suppose we perform another experiment involving the same reaction, only this time we begin the experiment with 500 molecules of reactant. This time, at the point of equilibrium, there will be 400 molecules of product present and 100 molecules of reactant remaining (400/100 = 4). As we began the two experiments with different numbers of reactant molecules, it's not troubling that there are different absolute numbers of product and reactant molecules present at equilibrium; however, the ratio, K, is the same for both experiments. We say that these two experiments represent two different positions of equilibrium: an equilibrium position corresponds to a particular set of experimental equilibrium concentrations that fulfill the value of the equilibrium constant. Any experiment that is performed with a different amount of starting material will come to its own unique equilibrium position, but the equilibrium constant ratio, K, will be the same for a given reaction regardless of the starting amounts taken.

14. Your paraphrase of Le Châtelier's principle should go something like this: "when you make any change to a system in equilibrium, this throws the system temporarily out of equilibrium, and the system responds by reacting in whichever direction will be able to reach a new position of equilibrium". There are various changes that can be made to a system in equilibrium. Following are examples:

a.   the concentration of one of the reactants is increased.

Consider the reaction: $2SO_2(g) + O_2(g) \rightleftarrows 2SO_3(g)$

Suppose the reactants have already reacted and a position of equilibrium has been reached that fulfills the value of K for the reaction. At this point there will be present particular amounts of each reactant and of product. Suppose then one additional mole of O_2 is added to the system from outside. At the instant the additional O_2 is added, the system will not be in equilibrium: there will be too much O_2 present in the system to be compatible with the amounts of SO_2 and SO_3 present. The system will respond by reacting to get rid of some of the excess O_2 until the value of the ratio K is again fulfilled. If the system reacts to get rid of the excess of O_2, additional product SO_3 will form. The net result is more SO_3 produced than if the the change had not been made.

b.   The concentration of one of the products is decreased by selectively removing it from the system.

Consider the reaction: $CH_3COOH + CH_3OH \rightleftarrows H_2O + CH_3COOCH_3$

This reaction is typical of many reactions involving organic chemical substances, in which two organic molecules react to form a larger molecule, with a molecule of water split out during the combination. This type of reaction on its own tends to come to equilibrium with only part of the starting materials being converted to the desired organic product (which effectively would leave the experimenter with a mixture of materials). A technique that is used by organic chemists to increase the effective yield of the desired organic product is to *separate* the two products (if the products are separated, they cannot react to give back the reactants). One method used is to add a drying agent to the mixture: such a drying agent chemically or physically absorbs the water from the system, removing it from equilibrium. If the water is removed, the reverse reaction cannot take place, and the reaction proceeds to a greater extent in the forward direction than if the drying agent had not been added. In other situations, an experimenter may separate the products of the reaction by distillation (if the boiling points make this possible): again, if the products have been separated, then the reverse reaction will not be possible, and the forward reaction will occur to a greater extent.

c.   The reaction system is compressed to a smaller volume.

Consider the example: $3H_2(g) + N_2(g) \rightleftarrows 2NH_3(g)$

For equilibria involving gases, when the volume of the reaction system is compressed suddenly, the pressure in the system increases. However, if the reacting system can relieve some of this increased pressure by reacting, it will do so. This will happen by the reaction occurring in whichever direction will give the smaller number of moles of gas (if the number of moles of gas is decreased in a particular volume, the pressure will decrease).

   For the reaction above, there are two moles of the gas on the right side of the equation, but there is a total of four moles on the left side. If this system at equilibrium were to be suddenly compressed to a smaller volume, the reaction would proceed further to the right (in favor of more ammonia being produced).

d.   The temperature is increased for an endothermic reaction.

   Consider the reaction: $2NaHCO_3 + heat \rightleftarrows Na_2CO_3 + H_2O + CO_2$

   Although a change in temperature actually does change the *value* of the equilibrium constant, we can simplify reactions involving temperature changes by treating heat energy as if it were a chemical substance: for this endothermic reaction, heat is one of the reactants. As we saw in the example in part (a) of this question, increasing the concentration of one of the reactants for a system at equilibrium causes the reaction to proceed further to the right, forming additional product. Similarly for the endothermic reaction given above, increasing the temperature causes the reaction to proceed further in the direction of products than if no change had been made. It is as if there were too much "heat" to be compatible with the amount of substances present. The substances react to get rid of some of the energy.

e.   the temperature is decreased for an exothermic process.

   Consider the reaction: $PCl_3 + Cl_2 \rightleftarrows PCl_5 + heat$

   As discussed in part (d) above, although changing the temperature at which a reaction is performed does change the numerical value of K, we can simplify our discussion of this reaction by treating heat energy as if it were a chemical substance. Heat is a product of this reaction. If we are going to lower the temperature of this reaction system, the only way to accomplish this is to remove energy from the system. Lowering the temperature of the system is really working with this system in its attempt to release heat energy. So lowering the temperature should favor the production of more product than if no change were made.

16.   Specific answer depends on student choice of examples. In general, for a weak acid, HA, and a weak base, B:

$HA + H_2O \rightleftarrows H_3O^+ + A^-$ $B + H_2O \rightleftarrows HB^+ + OH^-$

18.   a.   not a conjugate pair: they differ by more than one proton

   b.   not a conjugate pair: they differ by more than one proton

c.   a conjugate acid-base pair: H_3O^+ is the acid, H_2O is the base

d.   a conjugate acid-base pair: H_2O is the acid, OH^- is the base

e.   not a conjugate pair: they differ by more than one proton

20.   a. HNO_3 is a strong acid, so

$$[H^+] = 0.00141 \ M$$

$$pH = -\log(0.00141) = 2.851$$

$$pOH = 14 - 2.851 = 11.15$$

b.   NaOH is a strong base, so

$$[OH^-] = 2.13 \times 10^{-3} \ M$$

$$pOH = -\log(2.13 \times 10^{-3}) = 2.672$$

$$pH = 14 - 2.672 = 11.33$$

c.   HCl is a strong acid, so

$$[H^+] = 0.00515 \ M$$

$$pH = -\log(0.00515) = 2.288$$

$$pOH = 14 - 2.288 = 11.71$$

d. $Ca(OH)_2$ is a strong, but not very soluble base. Each formula unit of $Ca(OH)_2$ produces two formula units of OH^- ion.

$$[OH^-] = 2 \times 5.65 \times 10^{-5} \ M = 1.13 \times 10^{-4} \ M$$

$$pOH = -\log(1.13 \times 10^{-4}) = 3.947$$

$$pH = 14 - 3.947 = 10.05$$

22. $2SO_2(g) + O_2(g) \rightleftarrows 2SO_3(g)$

$$K = \frac{[SO_3]^2}{[SO_2]^2[O_2]} = \frac{[0.42]^2}{[1.4 \times 10^{-3}]^2[4.5 \times 10^{-4}]} = 2.0 \times 10^8$$

Chapter 17   Oxidation-Reduction Reactions/Electrochemistry

2.   Oxidation can be defined as the loss of electrons by an atom, molecule, or ion. Oxidation may also be defined as an increase in oxidation state for an element, but because elements can only increase their oxidation states by losing electrons, the two definitions are equivalent. The following equation shows the oxidation of copper metal to copper(II) ion

$$Cu \rightarrow Cu^{2+} + 2e^-$$

Reduction can be defined as the gaining of electrons by an atom, molecule, or ion. Reduction may also be defined as a decrease in oxidation state for an element, but naturally such a decrease takes place by the gaining of electrons (so the two definitions are equivalent). The following equation shows the reduction of sulfur atoms to sulfide ion.

$$S + 2e^- \rightarrow S^{2-}$$

4.   Each of these reactions involves a *metallic* element in the form of the *free* element on one side of the equation; on the other side of the equation, the metallic element is *combined* in an ionic compound. If a metallic element goes from the free metal to the ionic form, the metal is oxidized (loses electrons).

   a.   sodium is oxidized, nitrogen is reduced

   b.   magnesium is oxidized, chlorine is reduced

   c.   aluminum is oxidized, bromine is reduced

   d.   iron is oxidized, oxygen is reduced

6.   Each of these equations involve free elements on the left side, and the elements combined in a compound on the right side. When free elements react to form a compound, there is a change in oxidation state that indicates that an oxidation-reduction process has taken place.

   a.   potassium is oxidized, oxygen is reduced

   b.   hydrogen is oxidized, sulfur is reduced

   c.   hydrogen is oxidized, nitrogen is reduced

   d.   mercury is oxidized, chlorine is reduced

8.   The oxidation state of a *pure element* is *zero*, regardless of whether the element occurs naturally as single atoms or as a molecule.

10.   Fluorine is always assigned a negative oxidation state (–1) because all other elements are less electronegative. The other halogens are *usually* assigned an oxidation state of –1 in compounds. In interhalogen compounds such as ClF, fluorine is assigned oxidation state –1 (F is more electronegative than Cl). Chlorine, therefore, must be assigned a +1 oxidation state in this instance.

12.   Oxidation states represent a bookkeeping method to assign electrons in a molecule or ion. As an ion has a net charge, the sum of the oxidation states of the atoms in the ion must equal the charge on the ion. For example, the hydroxide ion (OH^-) has an overall charge of -1 because hydrogen has an oxidation state of $+1$, whereas oxygen has an oxidation state of -2 in the hydroxide ion: $(-2) + (+1) = -1$.

14.   The rules for assigning oxidation states are given in Section 17.2 of the text. The rule that applies for each element in the following answers is given in parentheses after the element and its oxidation state.

a.   P, 0 (Rule 1)

b.   H, +1 (Rule 4); Cl, +1 (Rule 6); O, -2 (Rule 3)

c.   C, -4 (Rule 6); H, +1 (Rule 4)

d.   H, +1 (Rule 4); P, +5 (Rule 6); O -2 (Rule 3)

16.   The rules for assigning oxidation states are given in Section 17.2 of the text. The rule that applies for each element in the following answers is given in parentheses after the element and its oxidation state.

a.   Cl, 0 (Rule 1)

b.   Co, +2 (Rule 6); Cl -1 (Rule 2)

c.   Co, +3 (Rule 6); Cl -1 (Rule 2)

d.   C, +4 (Rule 6); O, -2 (Rule 3)

18.   The rules for assigning oxidation states are given in Section 17.2 of the text. The rule that applies for each element in the following answers is given in parentheses after the element and its oxidation state.

a.   H, +1 (Rule 4); S, +6 (Rule 6); O, -2 (Rule 3)

b.   Mn, +7 (Rule 7); O, -2 (Rule 3)

c.   N, +5 (Rule 7); O, -2 (Rule 3)

d.   K, +1 (Rule 2); P, +5 (Rule 6); O, -2 (Rule 3)

20.   The rules for assigning oxidation states are given in Section 17.2 of the text. The rule that applies for each element in the following answers is given in parentheses after the element and its oxidation state.

a.   Cu, +2 (Rules 2,6); Cl, -1 (Rules 2,5)

b.   Cr, +3 (Rules 2,6); Cl, -1 (Rules 2,5)

c.   H, +1 (Rule 4); O, −2 (Rule 3); Cr, +6 (Rule 7)

d.   Cr, +3 (Rules 2,6); O, −2 (Rules 2,3)

22.   The rules for assigning oxidation states are given in Section 17.2 of the text. The rule that applies for each element in the following answers is given in parentheses after the element and its oxidation state.

a.   Mn, +2 (Rule 6), Cl, −1 (Rule 2)

b.   H, +1 (Rule 4); Br, +5 (Rule 6), O, −2 (Rule 3)

c.   B, +3 (Rule 6); H, −1 (H is *more* electronegative than B: Rule 5)

d.   C, +4 (Rule 6); F, −1 (Rule 5)

24.   Electrons are negative; when an atom gains electrons, it gains one negative charge for each electron gained. For example, in the reduction reaction $Cl + e^- \rightarrow Cl^-$, the oxidation state of chlorine decreases from 0 to −1 as the electron is gained.

26.   Answer depends on student example chosen.

28.   An oxidizing agent oxidizes another species by gaining the electrons lost by the other species; therefore, an oxidizing agent itself decreases in oxidation state. A reducing agent increases its oxidation state when acting on another atom or molecule.

30.   a. $Zn(s) + 2HNO_3(g) \rightarrow Zn(NO_3)_2(aq) + H_2(g)$

Zn 0   H +1   Zn +2   H 0

N +5   N +5

O −2   O −2

zinc is oxidized; hydrogen is reduced

b. $H_2(g) + CuSO_4(aq) \rightarrow Cu(s) + H_2SO_4(aq)$

H 0   Cu +2   Cu 0   H +1

S +6   S +6

O −2   O −2

hydrogen is oxidized; copper is reduced

c. $N_2(g) + 3Br_2(l) \rightarrow 2NBr_3(g)$

N 0   Br 0   N −3

Br +1

N is more electronegative than Br

nitrogen is reduced; bromine is oxidized

d. $2KBr(aq) + Cl_2(g) \rightarrow 2KCl(aq) + Br_2(l)$

K +1   Cl 0   K +1   Br 0

Br −1   Cl −1

bromine is oxidized; chlorine is reduced

32.   a. $Fe_2O_3(s) + 3C(s) \rightarrow 2Fe(s) + 3CO(g)$

Fe +3   C 0   Fe 0   C +2

O −2   O −2

carbon is oxidized; iron is reduced

b. $Mg(s) + 2HBr(aq) \rightarrow MgBr_2(aq) + H_2(g)$

Mg 0   H +1   Mg +2   H 0

            Br −1   Br −1

magnesium is oxidized; hydrogen is reduced

c. $2CoS(s) + S(s) \rightarrow Co_2S_3(s)$

Co +2   S 0   Co +3

S −2   S −2

cobalt is oxidized; sulfur (in S) is reduced

d. $2Ag(s) + 2HNO_3(aq) \rightarrow 2AgNO_3(aq) + H_2(g)$

Ag 0   H +1   Ag +1   H 0

            N +5   N +5

            O −2   O −2

silver is oxidized; hydrogen is reduced

34.   Iron is reduced [+3 in $Fe_2O_3(s)$, 0 in $Fe(l)$]; carbon is oxidized [+2 in $CO(g)$, +4 in $CO_2(g)$]. $Fe_2O_3(s)$ is the oxidizing agent; $CO(g)$ is the reducing agent.

36.   Chlorine is reduced [0 in $Cl_2(g)$, −1 in $NaCl(s)$]; bromine is oxidized [−1 in $NaBr(aq)$, 0 in $Br_2(l)$]. $Cl_2(g)$ is the oxidizing agent; $NaBr(aq)$ is the reducing agent.

38.   Oxidation-reduction reactions are often more complicated than "regular" reactions; frequently the coefficients necessary to balance the number

of electrons transferred come out to be large numbers. We also have to make certain that we account for the electrons being transferred.

40. Under ordinary conditions it is impossible to have "free" electrons that are not part of some atom, ion, or molecule. For this reason, the total number of electrons lost by the species being oxidized must equal the total number of electrons gained by the species being reduced.

42. a. $N_2(g) \rightarrow N^{3-}(s)$

   Balance nitrogen: $N_2(g) \rightarrow 2N^{3-}(s)$

   Balance charge: $\mathbf{6e^-} + N_2(g) \rightarrow 2N^{3-}(s)$

   Balanced half-reaction: $6e^- + N_2(g) \rightarrow 2N^{3-}(s)$

   b. $O_2^{2-}(aq) \rightarrow O_2(g)$

   Balance charge: $O_2^{2-}(aq) \rightarrow O_2(g) + \mathbf{2e^-}$

   Balanced half-reaction: $O_2^{2-}(aq) \rightarrow O_2(g) + 2e^-$

   c. $Zn(s) \rightarrow Zn^{2+}(aq)$

   Balance charge: $Zn(s) \rightarrow Zn^{2+}(aq) + \mathbf{2e^-}$

   Balanced half-reaction: $Zn(s) \rightarrow Zn^{2+}(aq) + 2e^-$

   d. $F_2(g) \rightarrow F^-(aq)$

   Balance fluorine: $F_2(g) \rightarrow \mathbf{2F^-}(aq)$

   Balance charge: $\mathbf{2e^-} + F_2(g) \rightarrow 2F^-(aq)$

   Balanced half-reaction: $2e^- + F_2(g) \rightarrow 2F^-(aq)$

44. a. $HBr(aq) \rightarrow Br^-(aq) + H_2(g)$

   Balance hydrogen: $\mathbf{2}HBr(aq) \rightarrow Br^-(aq) + H_2(g)$

   Balance bromine: $2HBr(aq) \rightarrow \mathbf{2}Br^-(aq) + H_2(g)$

   Balance charge: $\mathbf{2e^-} + 2HBr(aq) \rightarrow \mathbf{2}Br^-(aq) + H_2(g)$

   Balanced half-reaction: $2e^- + 2HBr(aq) \rightarrow \mathbf{2}Br^-(aq) + H_2(g)$

   b. $ClO_4^-(aq) \rightarrow ClO_3^-(aq)$

   Balance oxygen: $ClO_4^-(aq) \rightarrow ClO_3^-(aq) + \mathbf{H_2O}(l)$

   Balance hydrogen: $\mathbf{2H^+} + ClO_4^-(aq) \rightarrow ClO_3^-(aq) + H_2O(l)$

   Balance charge: $2e^- + \mathbf{2H^+} + ClO_4^-(aq) \rightarrow ClO_3^-(aq) + H_2O(l)$

   Balanced half-reaction: $\mathbf{2e^-} + \mathbf{2H^+} + ClO_4^-(aq) \rightarrow ClO_3^-(aq) + H_2O(l)$

c. $PbSO_4(s) \rightarrow Pb(s) + HSO_4^-$

Balance hydrogen: $\mathbf{H^+}(aq) + PbSO_4(s) \rightarrow Pb(s) + HSO_4^-$

Balance charge: $\mathbf{2e^-} + H^+(aq) + PbSO_4(s) \rightarrow Pb(s) + HSO_4^-$

Balanced half-reaction: $2e^- + H^+(aq) + PbSO_4(s) \rightarrow Pb(s) + HSO_4^-$

d. $HNO_2(aq) \rightarrow NO(g)$

Balance oxygen: $HNO_2(aq) \rightarrow NO(g) + \mathbf{H_2O}(l)$

Balance hydrogen: $\mathbf{H^+}(aq) + HNO_2(aq) \rightarrow NO(g) + \mathbf{H_2O}(l)$

Balance charge: $\mathbf{e^-} + H^+(aq) + HNO_2(aq) \rightarrow NO(g) + \mathbf{H_2O}(l)$

Balanced half-reaction: $e^- + H^+(aq) + HNO_2(aq) \rightarrow NO(g) + \mathbf{H_2O}(l)$

46.   a. $Al(s) + H^+(aq) \rightarrow Al^{3+}(aq) + H_2(g)$

$Al \rightarrow Al^{3+}$

Balance charge: $Al \rightarrow Al^{3+} + \mathbf{3e^-}$

$H^+ \rightarrow H_2$

Balance hydrogen: $\mathbf{2H^+} \rightarrow H_2$

Balance charge: $\mathbf{2e^-} + 2H^+ \rightarrow H_2$

Combine half-reactions: $3 \times (\mathbf{2e^-} + 2H^+ \rightarrow H_2)$

$2 \times (Al \rightarrow Al^{3+} + \mathbf{3e^-})$

$2Al(s) + 6H^+(aq) \rightarrow 2Al^{3+}(aq) + 3H_2(g)$

b. $S^{2-}(aq) + NO_3^-(g) \rightarrow S(s) + NO(g)$

$S^{2-} \rightarrow S$

Balance charge: $S^{2-} \rightarrow S + \mathbf{2e^-}$

$NO_3^- \rightarrow NO$

Balance oxygen: $NO_3^- \rightarrow NO + 2H_2O$

Balance hydrogen: $\mathbf{4H^+} + NO_3^- \rightarrow NO + 2H_2O$

Balance charge: $\mathbf{3e^-} + 4H^+ + NO_3^- \rightarrow NO + 2H_2O$

Combine half-reactions: $3 \times (S^{2-} \rightarrow S + 2e^-)$

$2 \times (3e^- + 4H^+ + NO_3^- \rightarrow NO + 2H_2O)$

$8H^+ + 3S^{2-}(aq) + 2NO_3^-(g) \rightarrow 3S(s) + 2NO(g) + 4H_2O$

c. $I_2(aq) + Cl_2(aq) \rightarrow IO_3^-(aq) + HCl(g)$

$I_2 \rightarrow IO_3^-$

Balance iodine: $I_2 \rightarrow \mathbf{2}IO_3^-$

Balance oxygen: $\mathbf{6H_2O} + I_2 \rightarrow IO_3^-$

Balance hydrogen: $6H_2O + I_2 \rightarrow IO_3^- + \mathbf{12H^+}$

Balance charge: $6H_2O + I_2 \rightarrow IO_3^- + 12H^+ + \mathbf{10e^-}$

$Cl_2 \rightarrow HCl$

Balance chlorine: $Cl_2 \rightarrow \mathbf{2}HCl$

Balance hydrogen: $\mathbf{2H^+} + Cl_2 \rightarrow 2HCl$

Balance charge: $\mathbf{2e^-} + 2H^+ + Cl_2 \rightarrow 2HCl$

Combine half-reactions: $5 \times (2e^- + 2H^+ + Cl_2 \rightarrow 2HCl)$

$6H_2O + I_2 \rightarrow IO_3^- + 12H^+ + 10e^-$

$6H_2O(l) + 2I_2(aq) + 5Cl_2(aq) \rightarrow 2IO_3^-(aq) + 10HCl(g) + 2H^+(aq)$

d. $AsO_4^-(aq) + S^{2-}(aq) \rightarrow AsO_3^-(s) + S(s)$

$AsO_4^- \rightarrow AsO_3^-$

Balance oxygen: $AsO_4^- \rightarrow AsO_3^- + \mathbf{H_2O}$

Balance hydrogen: $\mathbf{2H^+} + AsO_4^- \rightarrow AsO_3^- + H_2O$

Balance charge: $\mathbf{2e^-} + 2H^+ + AsO_4^- \rightarrow AsO_3^- + H_2O$

$S^{2-} \rightarrow S$

Balance charge: $S^{2-} \rightarrow S + \mathbf{2e^-}$

$2H^+(aq)\ AsO_4^-(aq) + S^{2-}(aq) \rightarrow AsO_3^-(s) + S(s) + H_2O(l)$

48. $Cl_2(g) + 2I^-(aq) \rightarrow 2Cl^-(aq) + I_2(s)$

$Cl_2(g) + 2Br^-(aq) \rightarrow 2Cl^-(aq) + Br_2(l)$

50.   A salt bridge typically consists of a *U*-shaped tube filled with an inert electrolyte (one involving ions that are not part of the oxidation -reduction reaction). A salt bridge is used to complete the electrical circuit in a cell. Any method that allows transfer of charge without allowing bulk mixing of the solutions may be used (another common method is to set up one half-cell in a porous cup, which is then placed in the beaker containing the second half-cell).

52.   In a galvanic cell, the anode is the electrode where oxidation occurs; the cathode is the electrode where reduction occurs.

54.

Pb^{2+}(aq) ion is reduced; Zn(s) is oxidized.

The reaction at the anode is Zn(s) → Zn^{2+}(aq) + 2e$^-$.

The reaction at the cathode is Pb^{2+}(aq) + 2e$^-$ → Pb(s)

56.   Both normal and alkaline cells contain zinc as one electrode; zinc corrodes more slowly under alkaline conditions than in the highly acidic environment of a normal dry cell.

anode: Zn(s) + 2OH$^-$(aq) → ZnO(s) + H$_2$O(l) + 2e$^-$

cathode: 2MnO$_2$(s) + H$_2$O(l) + 2e$^-$ → Mn$_2$O$_3$(s) + 2OH$^-$(aq)

58.   Aluminum is a very reactive metal when freshly isolated in the pure state. However, on standing for even a relatively short period of time, aluminum metal forms a thin coating of Al$_2$O$_3$ on its surface from reaction with atmospheric oxygen. This coating of Al$_2$O$_3$ is much less reactive than the metal and serves to protect the surface of the metal from further attack.

60.   In cathodic protection of steel tanks and pipes, a more reactive metal than iron is connected to the item to be protected. The active metal is then preferentially oxidized rather than the iron of the tank or pipe.

62.   The main recharging reaction for the lead storage battery is

2PbSO$_4$(s) + 2H$_2$O(l) → Pb(s) + PbO$_2$(s) + 2H$_2$SO$_4$(aq).

A major side reaction is the electrolysis of water

2H$_2$O(l) → 2H$_2$(g) + O$_2$(g).

This results in production of an explosive mixture of hydrogen and oxygen, which accounts for many accidents in recharging of such batteries.

64. Electrolysis is applied in electroplating by making the item to be plated the cathode in a cell containing a solution of ions of the desired plating metal.

66. loss; oxidation state

68. electronegative

70. An *oxidizing agent* is an atom, molecule, or ion that causes the oxidation of another species. During this process, the oxidizing agent is reduced.

72. lose

74. separate from

76. oxidation

78. An electrolysis reaction results when an electrical current from an outside source is used to cause an otherwise nonspontaneous reaction to occur. An example is the electrolysis of water: $2H_2O(l) \rightarrow 2H_2(g) + O_2(g)$; this reaction only takes place if an electrical current of sufficient voltage is passed through the water.

80. hydrogen; oxygen

82. oxidation

84. a. $4Fe(s) + 3O_2(g) \rightarrow 2Fe_2O_3(s)$

    iron is oxidized; oxygen is reduced

    b. $2Al(s) + 3Cl_2(g) \rightarrow 2AlCl_3(s)$

    aluminum is oxidized; chlorine is reduced

    c. $6Mg(s) + P_4(s) \rightarrow 2Mg_3P_2(s)$

    magnesium is oxidized; phosphorus is reduced

86. a.   aluminum is oxidized; hydrogen is reduced
    b.   hydrogen is reduced; iodine is oxidized
    c.   copper is oxidized; hydrogen is reduced

88. a. $C_3H_8(g) + 5O_2(g) \rightarrow 3CO_2(g) + 4H_2O(g)$
    b. $CO(g) + 2H_2(g) \rightarrow CH_3OH(l)$
    c. $SnO_2(s) + 2C(s) \rightarrow Sn(s) + 2CO(g)$
    d. $C_2H_5OH(l) + 3O_2(g) \rightarrow 2CO_2(g) + 3H_2O(g)$

90.   Each of these reactions involves a *metallic* element in the form of the
        free element on one side of the equation; on the other side of the equa-
        tion, the metallic element is *combined* in an ionic compound. If a metallic
        element goes from the free metal to the ionic form, the metal is oxidized
        (loses electrons).

        a.   sodium is oxidized; oxygen is reduced

        b.   iron is oxidized; hydrogen is reduced

        c.   oxygen (O^{2-}) is oxidized; aluminum (Al^{3+}) is reduced (this reaction
              is the reverse of the type discussed above)

        d.   magnesium is oxidized; nitrogen is reduced

92.   The rules for assigning oxidation states are given in Section 17.2 of the
        text. The rule that applies for each element in the following answers is
        given in parentheses after the element and its oxidation state.

        a.   H +1 (Rule 4); N −3 (Rule 6)

        b.   C +2 (Rule 6); O −2 (Rule 3)

        c.   C +4 (Rule 6); O −2 (Rule 3)

        d.   N +3 (Rule 6); F −1 (Rule 5)

94.   The rules for assigning oxidation states are given in Section 17.2 of the
        text. The rule that applies for each element is that given in parentheses
        after the element and its oxidation state.

        a.   Mn +4 (Rule 6); O −2 (Rule 3)

        b.   Ba +2 (Rule 2); Cr +6 (Rule 6); O −2 (Rule 3)

        c.   H +1 (Rule 4); S +4 (Rule 6); O −2 (Rule 3)

        d.   Ca +2 (Rule 2); P +5 (Rule 6); O −2 (Rule 3)

96.   The rules for assigning oxidation states are given in Section 17.2 of the
        text. The rule that applies for each element is that given in parentheses
        after the element and its oxidation state.

        a.   Bi +3 (Rule 7); O −2 (Rule 3)

        b.   P +5 (Rule 7); O −2 (Rule 3)

        c.   N +3 (Rule 7); O −2 (Rule 3)

        d.   Hg +1 (Rule 7)

98. a. $2B_2O_3(s) + 6Cl_2(g) \rightarrow 4BCl_3(l) + 3O_2(g)$

B +3   Cl 0   B +3   O 0

O -2   Cl -1

oxygen is oxidized; chlorine is reduced

b. $GeH_4(g) + O_2(g) \rightarrow Ge(s) + 2H_2O(g)$

Ge -4   O 0   Ge 0   H +1

H +1   O -2

germanium is oxidized; oxygen is reduced

c. $C_2H_4(g) + Cl_2(g) \rightarrow C_2H_4Cl_2(l)$

C -2   Cl 0   C -1

H +1   H +1; Cl -1

carbon is oxidized; chlorine is reduced

d. $O_2(g) + 2F_2(g) \rightarrow 2OF_2(g)$

O 0   F 0   O +2

                    F -1

oxygen is oxidized; fluorine is reduced

100. a. $SiO_2(s) \rightarrow Si(s)$

Balance oxygen: $SiO_2(s) \rightarrow Si(s) + \mathbf{2H_2O}(l)$

Balance hydrogen: $SiO_2(s) + \mathbf{4H^+}(aq) \rightarrow Si(s) + 2H_2O(l)$

Balance charge: $SiO_2(s) + 4H^+(aq) + \mathbf{4e^-} \rightarrow Si(s) + 2H_2O(l)$

Balanced half-reaction: $SiO_2(s) + 4H^+(aq) + 4e^- \rightarrow Si(s) + 2H_2O(l)$

b. $S(s) \rightarrow H_2S(g)$

Balance hydrogen: $S(s) + \mathbf{2H^+}(aq) \rightarrow H_2S(g)$

Balance charge: $S(s) + 2H^+(aq) + \mathbf{2e^-} \rightarrow H_2S(g)$

Balanced half-reaction: $S(s) + 2H^+(aq) + 2e^- \rightarrow H_2S(g)$

c. $NO_3^-(aq) \rightarrow HNO_2(aq)$

Balance oxygen: $NO_3^-(aq) \rightarrow HNO_2(aq) + \mathbf{H_2O}(l)$

Balance hydrogen: $NO_3^-(aq) + \mathbf{3H^+}(aq) \rightarrow HNO_2(aq) + H_2O(l)$

Balance charge: $NO_3^-(aq) + 3H^+(aq) + \mathbf{2e^-} \rightarrow HNO_2(aq) + H_2O(l)$

Balanced half-reaction: $NO_3^-(aq) + 3H^+(aq) + 2e^- \rightarrow HNO_2(aq) + H_2O(l)$

d. $NO_3^-(aq) \rightarrow NO(g)$

Balance oxygen: $NO_3^-(aq) \rightarrow NO(g) + \mathbf{2H_2O}(l)$

Balance hydrogen: $NO_3^-(aq) + \mathbf{4H^+}(aq) \rightarrow NO(g) + 2H_2O(l)$

Balance charge: $NO_3^-(aq) + 4H^+(aq) + \mathbf{3e^-} \rightarrow NO(g) + 2H_2O(l)$

Balanced half-reaction: $NO_3^-(aq) + 4H^+(aq) + 3e^- \rightarrow NO(g) + 2H_2O(l)$

102.   For simplicity, the physical states of the substances have been omitted until the final balanced equation is given.

For the reduction of the permanganate ion, MnO_4^-, in acid solution, the half-reaction is always the *same*:

$MnO_4^- \rightarrow Mn^{2+}$

Balance oxygen: $MnO_4^- \rightarrow Mn^{2+} + \mathbf{4H_2O}$

Balance hydrogen: $\mathbf{8H^+} + MnO_4^- \rightarrow Mn^{2+} + 4H_2O$

Balance charge: $8H^+ + MnO_4^- + \mathbf{5e^-} \rightarrow Mn^{2+} + 4H_2O$

a. $C_2O_4^{2-} \rightarrow CO_2$

Balance carbon: $C_2O_4^{2-} \rightarrow \mathbf{2}CO_2$

Balance charge: $C_2O_4^{2-} \rightarrow \mathbf{2}CO_2 + \mathbf{2e^-}$

Combine half-reactions: $5 \times (C_2O_4^{2-} \rightarrow \mathbf{2}CO_2 + 2e^-)$

$2 \times (8H^+ + MnO_4^- + 5e^- \rightarrow Mn^{2+} + 4H_2O)$

$16H^+(aq) + 2MnO_4^-(aq) + 5C_2O_4^{2-}(aq) \rightarrow 2Mn^{2+}(aq) + 8H_2O(l) + 10CO_2(g)$

b. $Fe^{2+} \rightarrow Fe^{3+}$

Balance charge: $Fe^{2+} \rightarrow Fe^{3+} + \mathbf{e^-}$

Combine half-reactions: $5 \times (Fe^{2+} \rightarrow Fe^{3+} + e^-)$

$8H^+ + MnO_4^- + 5e^- \rightarrow Mn^{2+} + 4H_2O$

$8H^+(aq) + MnO_4^-(aq) + 5Fe^{2+}(aq) \rightarrow Mn^{2+}(aq) + 4H_2O(l) + 5Fe^{3+}(aq)$

c. $Cl^- \rightarrow Cl_2$

Balance chlorine: $\mathbf{2}Cl^- \rightarrow Cl_2$

Balance charge: $2Cl^- \rightarrow Cl_2 + \mathbf{2e^-}$

Combine half-reactions: $5 \times (2Cl^- \rightarrow Cl_2 + 2e^-)$

$2 \times (8H^+ + MnO_4^- + 5e^- \rightarrow Mn^{2+} + 4H_2O)$

$16H^+(aq) + 2MnO_4^-(aq) + 10Cl^-(aq) \rightarrow 2Mn^{2+}(aq) + 8H_2O(l) + 5Cl_2(g)$

Chapter 18 Radioactivity and Nuclear Energy

2. The radius of a typical atomic nucleus is on the order of 10^{-13} cm, which is about one hundred thousand times smaller than the radius of an atom overall.

4. False. The mass number (A) of a nucleus represents the *total number* of protons *and* neutrons in the nucleus.

6. The atomic number (Z) is written in such formulas as a left subscript, whereas the mass number (A) is written as a left superscript. That is, the general symbol for a nuclide is $_Z^A X$. As an example, consider the isotope of oxygen with 8 protons and 8 neutrons: its symbol would be $_8^{16}O$.

8. neutron: charge 0, mass number 1, symbol $_0^1 n$

10. When a nucleus produces a beta (β) particle, the atomic number of the parent nucleus is *increased* by *one* unit. The beta particle has a mass number of zero, but an "atomic number" of –1.

12. Gamma rays are high energy photons of electromagnetic radiation. Gamma rays are not considered to be particles. When a nucleus produces only gamma radiation, the atomic number and mass number of the nucleus do not change. Gamma rays represent the energy changes associated with transitions and rearrangement of the particles within the nucleus.

14. Electron capture occurs when one of the inner orbital electrons is pulled into and becomes part of the nucleus.

16. The average atomic mass listed on the Periodic Table is 20.18, which suggests that the isotope of mass number 20 predominates in naturally occurring neon. The average atomic mass given in the periodic table is a *weighted* average that includes not only the masses of the isotopes, but also their relative abundances.

18. The approximate atomic molar mass could be calculated as follows: $0.79(24) + 0.10(25) + 0.11(26) = 24.3$. This is *only* an approximation because the mass numbers, rather than the actual isotopic masses, were used. The fact that the approximate mass calculated is slightly above 24 shows that the isotope of mass number 24 predominates.

20. a. $_{-1}^0 e$ or $_{-1}^0 \beta$

    b. $_{+1}^0 e$ or $_{+1}^0 \beta$

    c. $_0^0 \gamma$

22. a. $_{83}^{192} Bi$

b. $^{204}_{82}\text{Pb}$

c. $^{206}_{84}\text{Po}$

24.   a. $^{218}_{86}\text{Rn}$

b. $^{0}_{+1}\text{e}$ (positron)

c. $^{137}_{56}\text{Ba}$

26.   a. $^{136}_{53}\text{I} \rightarrow {}^{136}_{54}\text{Xe} + {}^{0}_{-1}\text{e}$

b. $^{133}_{51}\text{Sb} \rightarrow {}^{133}_{52}\text{Te} + {}^{0}_{-1}\text{e}$

c. $^{117}_{49}\text{In} \rightarrow {}^{117}_{50}\text{Sn} + {}^{0}_{-1}\text{e}$

d. $^{47}_{20}\text{Ca} \rightarrow {}^{47}_{21}\text{Sc} + {}^{0}_{-1}\text{e}$

28. $^{234}_{92}\text{U} \rightarrow {}^{230}_{90}\text{Th} \rightarrow {}^{226}_{88}\text{Ra} \rightarrow {}^{222}_{86}\text{Rn} \rightarrow {}^{218}_{84}\text{Po} \rightarrow {}^{214}_{82}\text{Pb}$

30.   In a nuclear bombardment process, a target nucleus is bombarded with high-energy particles (typically subatomic particles or small atoms) from a particle accelerator. This may result in the transmutation of the target nucleus into some other element. For example, nitrogen-14 may be transmuted into oxygen-17 by bombardment with alpha particles. There is often considerable repulsion between the target nucleus and the particles being used for bombardment (especially if the bombarding particle is positively charged like the target nucleus). Using accelerators to increases the kinetic energy of the bombarding particles can overcome this repulsion.

32. $^{27}_{13}\text{Al} + {}^{4}_{2}\text{He} \rightarrow {}^{30}_{15}\text{P} + {}^{1}_{0}\text{n}$

34.   The half-life of a nucleus is the time required for one-half of the original sample of nuclei to decay. A given isotope of an element always has the same half-life, although different isotopes of the same element may have greatly different half-lives. Nuclei of different elements typically have different half-lives.

36. $^{226}_{88}Ra$ is the most stable (longest half-life)

$^{224}_{88}Ra$ is the "hottest" (shortest half-life)

38.   highest activity   lowest activity

$^{87}Sr > ^{99}Tc > ^{24}Na > ^{99}Mo > ^{133}Xe > ^{131}I > ^{32}P > ^{51}Cr > ^{59}Fe$

40.   For ^{223}Ra, the half-life is 12 days. After two half-lives (24 days), 250 mg remains; after three half-lives (36 days), 125 mg remains.

For ^{224}Ra, the half-life is 3.6 days. One month would be approximately 8 half-life periods (29 days), and approximately 4 mg remains.
For ^{225}Ra, the half-life is 15 days. One month would be two half-life periods, and 250 mg remains.

42.   For an administered dose of 100 μg, 0.39 μg remains after 2 days. The fraction remaining is 0.39/100 = 0.0039; on a percentage basis, less than 0.4% of the original radioisotope remains.

44.   Carbon-14 is produced in the upper atmosphere by the bombardment of ordinary nitrogen with neutrons from space:

$$^{14}_{7}N + ^{1}_{0}n \rightarrow ^{14}_{6}C + ^{1}_{1}H$$

46.   We assume that the concentration of C-14 in the atmosphere is effectively constant. A living organism is constantly replenishing C-14 either through the processes of metabolism (sugars ingested in foods contain C-14), or photosynthesis (carbon dioxide contains C-14). When a plant dies, it can no longer replenish, and as the C-14 undergoes radioactive decay, its amount decreases with time.

48.   These isotopes and their uses are listed in Table 18.4. Some important examples include the use of I-131 (and other iodine isotopes) in the diagnosis and treatment of thyroid disease (iodine is used in the body primarily in the thyroid gland); Fe-59 in the study of the function of red blood cells (iron is a constituent of hemoglobin that is found in the red blood cells); Sr-87 in the study of bones (Sr is a Group 2 element, and is able to take the place of Ca in bone structures).

50.   Combining two light nuclei to form a heavier, more stable nucleus is called nuclear *fusion*. Splitting a heavy nucleus into nuclei with smaller mass numbers is called nuclear *fission*.

52. $^{1}_{0}n + ^{235}_{92}U \rightarrow ^{142}_{56}Ba + ^{91}_{36}Kr + 3^{1}_{0}n$ is one possibility.

54. A critical mass of a fissionable material is the amount needed to provide a high enough internal neutron flux to sustain the chain reaction (enough neutrons are produced to cause the continuous fission of further material). A sample with less than a critical mass is still radioactive, but cannot sustain a chain reaction.

56. An actual nuclear explosion, of the type produced by a nuclear weapon, cannot occur in a nuclear reactor because the concentration of the fissionable materials is not sufficient to form a supercritical mass. However, since many reactors are cooled by water, which can decompose into hydrogen and oxygen gases, a *chemical* explosion is possible that could scatter the radioactive material used in the reactor.

58. Breeder reactors use the more common uranium-238 isotope, which is bombarded with neutrons. Through a three-step process, the uranium-238 is converted to plutonium-239, which is fissionable and can then be used to fuel another reactor. The major controversy over breeder reactors concerns the extreme chemical toxicity of plutonium and the difficulty in handling it.

60. In one type of fusion reactor, two ^2_1H atoms are fused to produce ^4_2He. Because the hydrogen nuclei are positively charged, extremely high energies (temperatures of 40 million K) are needed to overcome the repulsion between the nuclei as they are shot into each other.

62. In the theory of stellar nucleosynthesis, it is considered that the nucleus began as a cloud of neutrons which exploded (the Big Bang). After this initial explosion, neutrons were thought to have decomposed into protons and electrons

$$^1_0n \rightarrow \, ^1_1H + \, ^0_{-1}e$$

The products of this decomposition were then thought to have combined to form large clouds of hydrogen atoms. As the hydrogen clouds became larger, gravitational forces caused these clouds to contract and heat up. Eventually the clouds of hydrogen were so dense and so hot that fusion of hydrogen nuclei into helium nuclei took place, with a great release of energy. When the tendency for the hydrogen clouds to expand from the heat of fusion was counter-balanced by the gravitational forces of the cloud, a small star had formed. In addition to the fusion of hydrogen nuclei into helium mentioned already, as the star's hydrogen supply is exhausted, the helium present in the star also begins to undergo fusion into nuclei of other elements.

64. Somatic damage is directly to the organism itself, causing nearly immediate sickness or death to the organism. Genetic damage is to the genetic machinery of the organism, which will be manifested in future generations of offspring.

66. Gamma rays penetrate long distances, but seldom cause ionization of biological molecules. Alpha particles, because they are much heavier although less penetrating, are very effective at ionizing biological

molecules and leave a dense trail of damage in the organism. Isotopes that release alpha particles can be ingested or breathed into the body where the damage from the alpha particles will be more acute.

68. Nuclear waste may remain radioactive for thousands of years, and much of it is chemically poisonous as well as radioactive. Most reactor waste is still in "temporary storage." Various suggestions have been made for a more permanent solution, such as casting the spent fuel into glass bricks to contain it, and then storing the bricks in corrosion-proof metal containers deep undergroud. No agreement on a permanent solution to the disposal of nuclear waste has yet been reached.

70. radioactive

72. mass

74. neutron; proton

76. radioactive decay

78. mass number

80. transuranium

82. half-life

84. radiotracers

86. chain

88. breeder

90. 4.5×10^9 dollars ($4.5 billion)

92. 3.5×10^{-11} J/atom; 8.9×10^{10} J/g

94. Despite the fact that nuclear waste has been generated for over 40 years, no permanent disposal plan has been implemented as yet. One proposal to dispose of such waste calls for the waste to be sealed in blocks of glass. These in turn are sealed in corrosion-proof metal drums, which would then be buried in deep, stable rock formations away from earthquake and other geologically active zones. In these deep storage areas, it is hoped that the waste could decay safely undisturbed until the radioactivity drops to "safe" levels.

96. $^{27}_{13}$Al: 13 protons, 14 neutrons

$^{28}_{13}$Al: 13 protons, 15 neutrons

29

$^{13}_{}$Al: 13 protons, 16 neutrons

98.   a. $^{0}_{-1}e$

b. $^{74}_{34}Se$

c. $^{240}_{92}U$

100.   For a decay of 10 μg to 1/1000 of this amount, we want to know when the amount of remaining ^{131}I is on the order of 0.01 μg.

time, days	0	8	16	24	32	40
mass, μg	10	5	2.5	1.25	0.625	0.313
time, days	48	56	64	72	80	
mass, μg	0.156	0.078	0.039	0.020	0.01	

Approximately 80 days are required.

102.   Breeder reactors are set up to convert non-fissionable ^{238}U into fissionable ^{239}Pu. The material used for fission in a breeder reactor is a combination of U-235 (which undergoes fission in a chain reaction) and the more common U-238 isotope. Excess neutrons from the U-235 fission are absorbed by the U-238 converting it to the fissionable plutonium isotope Pu-239. Although Pu-239 is fissionable, its chemical and physical properties make it very difficult and expensive to handle and process.

Chapter 19   Organic Chemistry

2.   A given carbon atom can be attached to a maximum of four other atoms.
      Carbon atoms have four valence electrons. By making four bonds, carbon
      atoms exactly complete their valence octet.

4.   A triple bond represents the sharing of *six* electrons (three *pairs* of
      electrons). The simplest example of an organic molecule containing a
      triple bond is acetylene, $H:C:::C:H$ ($H-C\equiv C-H$).

6.   Each carbon atom in ethane is bonded to *four* other atoms. According to
      VSEPR theory, each carbon atom has its electron pairs arranged
      tetrahedrally.

8.   An unsaturated hydrocarbon is one that contains an area of multiple
      bonding (double bond or triple bond) between some of the carbon atoms.
      Such a hydrocarbon is called "unsaturated" because each carbon atom of
      the multiple bond could bond to additional atoms (rather than to each
      other). Specific examples will depend on students' choice.

10.   Each successive member of this family differs from the previous member
      by a $-CH_2-$ unit (sometimes called a "methylene" unit). Such a family of
      compounds is often referred to as a *homologous series*.

12.   The general formula for the alkanes is C_nH_{2n+2}.

      a.   2(8) + 2 = 18 H atoms   b.   2(16) + 2 = 34 H atoms

      c.   2(1) + 2 = 4 H atoms   d.   2(4) + 2 = 10 H atoms

14.

a.
```
        H  H  H  H  H
        |  |  |  |  |
     H--C--C--C--C--C--H
        |  |  |  |  |
        H  H  H  H  H
```

pentane CH₃CH₂CH₂CH₂CH₃

b.
```
        H  H
        |  |
     H--C--C--H
        |  |
        H  H
```

ethane CH₃CH₃

c.
```
     H  H  H  H  H  H  H  H  H  H  H
     |  |  |  |  |  |  |  |  |  |  |
  H--C--C--C--C--C--C--C--C--C--C--C--H
     |  |  |  |  |  |  |  |  |  |  |
     H  H  H  H  H  H  H  H  H  H  H
```

undecane CH₃CH₂CH₂CH₂CH₂CH₂CH₂CH₂CH₂CH₂CH₃

d.
```
        H
        |
     H--C--H
        |
        H
```

methane CH₄

16.   branch or substituent

18.   With eight carbon atoms, there are a *lot* of isomers possible. Here are
      carbon skeletons for some of them:

20. The root name is derived from the number of carbon atoms in the *longest continuous chain* of carbon atoms.

22. The numbers indicate to which carbon of the longest continuous chain of carbon atoms the substituents are attached. The longest continuous chain is numbered from the end closest to the first substituent so as to give the lowest possible locator numbers.

24. Multiple substituents are listed in alphabetical order, disregarding any prefix.

26. a.   3-methylpentane

    b.   2,3,4,5-tetramethylhexane

    c.   3,4-diethyl-2,5-dimethylhexane (3-ethyl-2-methyl-4-isopropylhexane)

    d.   2-methylpentane

28. a.

$$CH_3-\underset{\underset{CH_3}{|}}{\overset{\overset{CH_3}{|}}{C}}-CH_2-CH_2-CH_2-CH_2-CH_3$$

    b.

$$CH_3-\underset{\underset{CH_3}{|}}{\overset{\overset{CH_3}{|}}{CH}}-CH_2-CH_2-CH_2-CH_3$$

    c.

$$CH_3-CH_2-\underset{\underset{CH_3}{|}}{\overset{\overset{CH_3}{|}}{C}}-CH_2-CH_2-CH_2-CH_3$$

    d.

$$CH_3-\underset{\underset{CH_3}{|}}{CH}-CH_2-\underset{\underset{CH_3}{|}}{CH}-CH_2-CH_2-CH_3$$

30.

Number of C atoms	Use
C_5-C_{12}	gasoline
$C_{10}-C_{18}$	kerosene, jet fuel
$C_{15}-C_{25}$	diesel fuel, heating oil, lubrication
$C_{25}-$	asphalt

32. Tetraethyl lead was added to gasolines to prevent "knocking" of high efficiency automobile engines. The use of tetraethyl lead is being phased

out because of the danger to the environment posed by the lead in this substance.

34.   Combustion represents the vigorous reaction of a hydrocarbon (or other substance) with oxygen. The combustion of alkanes has been used as a source of heat, light, and mechanical energy.

36.   This represents a dehydrogenation reaction. Dehydrogenation reactions involve the removal of hydrogen atoms from adjacent carbon atoms in an alkane (or other substance). When two hydrogen atoms are removed from an alkane or related compound, a double bond is created.

38.   a. $2C_6H_{14}(l) + 19O_2(g) \rightarrow 12CO_2(g) + 14H_2O(g)$

       b. $CH_4(g) + Cl_2(g) \rightarrow CH_3Cl(l) + HCl(g)$

       c. $CHCl_3(l) + Cl_2(g) \rightarrow CCl_4(l) + HCl(g)$

40.   An alkyne is a hydrocarbon containing a carbon-carbon triple bond. The general formula is C_nH_{2n-2}.

42.   The location of a double or triple bond in the longest chain of an alkene or alkyne is indicated by giving the *number* of the lowest number carbon atom involved in the double or triple bond.

44.   hydrogenation

46.   a.   3-bromo-1-butyne   b.   3-bromo-1-butene

       c.   3,4-dichloro-1-butene   d.   1,4-dichloro-2-butyne

48.   Shown are carbon skeletons:

```
C≡C-C-C-C-C        C-C≡C-C-C-C        C-C-C≡C-C-C          C≡C-C-C-C
                                              |                    |
                                              C                    C

C-C≡C-C-C          C≡C-C-C-C          C≡C-C-C
      |                  |                  |
      C                  C                  C
```

50.   For benzene, a *set* of equivalent Lewis structures can be drawn, differing only in the *location* of the three double bonds in the ring. Experimentally, however, benzene does not demonstrate the chemical properties expected for molecules having *any* double bonds. We say that the "extra" electrons that would go into making the second bond of the three double bonds are delocalized around the entire benzene ring; this delocalization of the electrons explains benzene's unique properties.

52. When named as a substituent, the benzene ring is called the *phenyl* group. Two examples are:

3-phenyl-1-butene   2-phenylhexane

54. *ortho-* refers to adjacent substituents (1,2-); *meta-* refers to two substituents with one unsubstituted carbon atom between them (1,3-); *para-* refers to two substituents with two unsubstituted carbon atoms between them (1,4-).

56.   a.   1,4-dibromobenzene (*p*-dibromobenzene)

       b.   naphthalene

       c.   3-chloro-1-methylbenzene (3-chlorotoluene)

       d.   1,2-dinitrobenzene (*o*-dinitrobenzene)

58.   a.   ether   b.   alcohol

       c.   alcohol   d.   organic (carboxylic) acid

60. Primary alcohols have *one* hydrocarbon fragment (alkyl group) bonded to the carbon atom where the –OH group is attached. Secondary alcohols have *two* such alkyl groups attached, and tertiary alcohols contain *three* such alkyl groups. Examples are:

ethanol (primary) CH_3-CH_2-OH

2-propanol (secondary) $CH_3-CH-CH_3$
                                           |
                                           OH

                                          CH_3
                                           |
2-methyl-2-propanol (tertiary) CH_3-C-CH_3
                                           |
                                           OH

62. Specific examples will depend on students' choice. Below are examples for alcohols with five carbon atoms.

$CH_3-CH_2-CH_2-CH_2-CH_2-OH$ primary   (1-pentanol)

$$CH_3-CH-CH_2-CH_2-CH_3$$
|
OH
   secondary   (2-pentanol)

OH
|
$$CH_3-C-CH_2-CH_3$$
|
CH_3
   tertiary   (2-methyl-2-butanol)

64. $C_6H_{12}O_6$ –yeast→ $2CH_3-CH_2-OH$ + $2CO_2$

The yeast necessary for the fermentation process are killed if the concentration of ethanol is over 13%. More concentrated ethanol solutions are most commonly made by distillation.

66.   methanol (CH_3OH) – starting material for synthesis of acetic acid and many plastics

ethylene glycol (CH_2OH-CH_2OH) – automobile antifreeze

isopropyl alcohol (2-propanol, $CH_3-CH(OH)-CH_3$) – rubbing alcohol

68.   Aldehydes and ketones both contain the carbonyl group $\diagdown C=O$. $\diagup$

Aldehydes and ketones differ in the *location* of the carbonyl function: aldehydes contain the carbonyl group at the end of a hydrocarbon chain (the carbon atom of the carbonyl group is bonded only to at most one other carbon atom); the carbonyl group of ketones represents one of the interior carbon atoms of a chain (the carbon atom of the carbonyl group is bonded to two other carbon atoms).

70.   Aldehydes and ketones are produced by the oxidation of primary and secondary alcohols, respectively.

CH_3-CH_2-OH ---oxidation--→ $CH_3-C=O$
|
H

$CH_3-CH-OH$ ---oxidation--→ $CH_3-C=O$
| |
CH_3 CH_3

72.   In addition to their systematic names (based on the hydrocarbon root, with the ending -*one*), ketones can also be named by naming the groups attached to either side of the carbonyl carbon as alkyl groups, followed by the word "ketone". Examples are:

$CH_3-C(=O)-CH_2CH_3$ methyl ethyl ketone (2-butanone, butanone)

$CH_3CH_2-C(=O)-CH_2CH_3$ diethyl ketone (3-pentanone)

74.   a.   CH₃-C-φ   [φ represents the phenyl group (benzene ring)]
            ‖
            O

      b.   CH₃-CH₂-CH₂-C=O
                        |
                        H

      c.   CH₃-CH₂-C-CH₃
                  ‖
                  O

      d.   CH₃-CH₂-CH₂-C-CH₂-CH₂-CH₃
                      ‖
                      O

      e.   CH₃-CH₂-CH₂-C-CH₂-CH₂-CH₃   (same molecule as above)
                      ‖
                      O

76.   Carboxylic acids are typically *weak* acids.

      $CH_3-CH_2-COOH(aq) \rightleftharpoons H^+(aq) + CH_3-CH_2-COO^-(aq)$

78. $CH_3-CH_2-CH_2-OH \rightarrow CH_3-CH_2-CHO$

      $CH_3-CH_2-CH_2-OH \rightarrow CH_3-CH_2-COOH$

80.   Acetylsalicylic acid is synthesized from salicylic acid (behaving as an alcohol through its -OH group) and acetic acid.

82.   a.   CH₃-CH₂-CH-CH₂-C=O
                     | |
                     CH₃   OH

      b.   H-C-O-CH₂-CH₃
              ‖
              O

      c.   φ-C-O-CH₃   [φ represents the pheynyl group (benzene ring)]
              ‖
              O

d.

$$CH_3-CH_2-\underset{\underset{Br}{|}}{CH}-\underset{\underset{OH}{|}}{C}=O$$

e.

$$CH_3-CH_2-\underset{\underset{CH_3}{|}}{CH}-\overset{\overset{Cl}{|}}{CH}-\underset{\underset{CH_3}{|}}{CH}-\overset{\overset{OH}{|}}{C}=O$$

84.   In addition polymerization, the monomer units simply add together to form the polymer, with no other products. Polyethylene and polytetrafluoroethylene (Teflon) are common examples.

86.   A polyester is formed from the reaction of a dialcohol (two –OH groups) with a diacid (two –COOH groups). One –OH group of the alcohol forms an *ester linkage* with one of the –COOH groups of the acid. As the resulting dimer still possesses an –OH and a –COOH group, the dimer can undergo further esterification reactions. Dacron is a common polyester.

88.

nylon

$$\left(\!\!-\underset{\underset{H}{|}}{N}-(CH_2)_6-\underset{\underset{H}{|}}{N}-\overset{\overset{O}{\|}}{C}-(CH_2)_4-\overset{\overset{O}{\|}}{C}-\!\!\right)$$

dacron

$$\left(\!\!-O-CH_2-CH_2-O-\overset{\overset{O}{\|}}{C}-\!\!\bigcirc\!\!-\overset{\overset{O}{\|}}{C}-\!\!\right)$$

90.   saturated

92.   straight-chain or normal

94.   -*ane*

96.   number

98.   anti-knocking

100.   substitution

102.   hydrogenation

104.   functional

106.   carbon monoxide

108.   carbonyl

110.   carboxyl

112.   addition

114.   The carbon skeletons are:

C-C-C-C-C .

C-C-C-C
|
C

C-C-C with C above and C below middle carbon

$$\begin{array}{c} C \\ | \\ C-C-C \\ | \\ C \end{array}$$

116.   a.   2-chlorobutane

b.   1,2-dibromoethane

c.   triiodomethane (common name: iodoform)

d.   2,3,4-trichloropentane

e.   2,2-dichloro-4-isopropylheptane

118.   a.
$$\begin{array}{c} CH_3 \\ | \\ CH_3-CH-CH-CH_2-CH_2-CH_2-CH_3 \\ | \\ CH_3 \end{array}$$

b.
$$\begin{array}{c} CH_3 \\ | \\ HO-CH_2-C-CH-CH_2-CH_2-CH_2-CH_2-CH_3 \\ |\ \ | \\ CH_3\ Cl \end{array}$$

c.
$$\begin{array}{c} CH_2=C-CH_2-CH_2-CH_2-CH_3 \\ | \\ Cl \end{array}$$

d.
$$\begin{array}{c} CH_2-CH=CH-CH_2-CH_2-CH_3 \\ | \\ Cl \end{array}$$

e.

120.   primary $CH_3-CH_2-CH_2-CH_2-CH_2-CH_2-OH$

secondary
$$\begin{array}{c} CH_3-CH_2-CH_2-CH_2-CH-CH_3 \\ | \\ OH \end{array}$$

tertiary
$$\begin{array}{c} CH_3 \\ | \\ CH_3-CH_2-C-CH_2-CH_3 \\ | \\ OH \end{array}$$

122.

CH₂-CH-CH-CH-CH-C=O
| | | | | |
OH OH OH OH OH H

124. a.

CH₃-C-CH₂-CH₂-CH₂-CH₂-CH₃
‖
O

b.

CH₃-CH₂-CH-CH₂-C=O
| |
CH₃   H

c.

CH₃-CH₂-CH₂-CH-CH₂-OH
|
CH₃

d.

OH
|
CH₂-CH-CH₂
| |
OH   OH

e.

CH₃
|
CH₃-CH-C-CH₂-CH₂-CH₃
‖
O

126.

HO—C=O   HO—C=O
| |
CH₃-CH-N-H + HO—C=O → CH₃-CH-N-C=O + H₂O
| | | |
H   CH₂—NH₂   H   CH₂—NH₂

128 a.

H H H H H H H H
| | | | | | | |
H-C-C-C-C-C-C-C-C-H
| | | | | | | |
H H H H H H H H

octane CH₃CH₂CH₂CH₂CH₂CH₂CH₂CH₃

b.

H H H H H H
| | | | | |
H-C-C-C-C-C-C-H
| | | | | |
H H H H H H

hexane CH₃CH₂CH₂CH₂CH₂CH₃

c.

$$
\begin{array}{c}
\text{H H H H} \\
| \ | \ | \ | \\
\text{H–C–C–C–C–H} \\
| \ | \ | \ | \\
\text{H H H H}
\end{array}
$$

butane $CH_3CH_2CH_2CH_3$

d.

$$
\begin{array}{c}
\text{H H H H H} \\
| \ | \ | \ | \ | \\
\text{H–C–C–C–C–C–H} \\
| \ | \ | \ | \ | \\
\text{H H H H H}
\end{array}
$$

pentane $CH_3CH_2CH_2CH_2CH_3$

130. a. 2,3-dimethylbutane

b. 3,3-diethylpentane

c. 2,3,3-trimethylhexane

d. 2,3,4,5,6-pentamethylheptane

132. a. $CH_3Cl(g)$

b. $H_2(g)$

c. $HCl(g)$

134. $CH≡C-CH_2-CH_2-CH_2-CH_2-CH_2-CH_3$ 1-octyne

$CH_3-C≡C-CH_2-CH_2-CH_2-CH_2-CH_3$ 2-octyne

$CH_3-CH_2-C≡C-CH_2-CH_2-CH_2-CH_3$ 3-octyne

$CH_3-CH_2-CH_2-C≡C-CH_2-CH_2-CH_3$ 4-octyne

136. a. carboxylic acid

b. ketone

c. ester

d. alcohol (phenol)

138. a. 3-methylpentanal

$$
\begin{array}{c}
CH_3\!-\!\!CH_2\!-\!\!CH\!-\!\!CH_2\!-\!\!C\!=\!\!O \\
\qquad\quad | \qquad\qquad | \\
\qquad\quad CH_3 \qquad\quad H
\end{array}
$$

b.   3-methyl-2-pentanone

$$CH_3-CH_2-\underset{\underset{CH_3}{|}}{CH}-\underset{\underset{O}{||}}{C}-CH_3$$

c.   methyl phenyl ketone

d.   2-hydroxybutanal

$$CH_3-CH_2-\underset{\underset{OH}{|}}{CH}-\underset{\underset{H}{|}}{C}=O$$

e.   propanal

$$CH_3-CH_2-\underset{\underset{H}{|}}{C}=O$$

140.   a. $CH_3-\underset{\underset{CH_3}{|}}{CH}-CH_2-COOH$

b.

c. $CH_3-CH_2-CH_2-CH_2-CH_2-COOH$

d. CH_3-COOH

Chapter 20   Biochemistry

2.   Proteins represent biopolymers of α-amino acids used for many purposes in the human body (structure, enzymes, antibodies, etc.). Proteins make up about 15% of the body by mass.

4.   Fibrous

6.   The structures of the amino acids are given in detail in Figure 20.2. Generally a side chain is nonpolar if it is mostly hydrocarbon in nature (e.g., alanine, in which the side chain is a methyl group). Side chains are polar if they contain the hydroxyl group (–OH), the sulfhydryl group (–SH), or a second amino (–NH$_2$) or carboxyl (–COOH) group.

8.   Figure 20.2 shows the amino acids separated into the hydrophilic and hydrophobic subgroupings. Note that the hydrophobic amino acids contain R groups that are basically hydrocarbon (or substituted hydrocarbon) in nature. This makes these *R* groups nonpolar and unlikely to interact with polar water molecules. Note, too, that the hydrophilic amino acids all contain a very polar functional group (e.g., –OH, –SH, –C=O, etc.) that enable the amino acid *R* groups to interact with water. For a protein in an aqueous medium, the hydrophilic *R* groups will orient themselves towards the aqueous medium, whereas the hydrophobic *R* groups will turn away from the aqueous medium, causing a profound effect on the protein's shape.

10.   There are six tripeptides possible.

cys-ala-phe   *ala-cys-phe*   *phe-ala-cys*

cys-phe-ala   *ala-phe-cys*   *phe-cys-ala*

12.   The "peptide linkage" is an amide linkage:

```
–C–N–
 ‖ |
 O H
```

14.   Long, thin, resilient proteins, such as hair, typically contain elongated, elastic alpha-helical protein molecules. Other proteins, such as silk, which in bulk form sheets or plates, typically contain protein molecules having the beta pleated sheet secondary structure. Proteins that do not have a structural function in the body, such as hemoglobin, typically have a globular structure.

16.   The secondary structure, in general, describes the arrangement of the long polypeptide chain of the protein. In the alpha-helical secondary structure, the chain forms a coil or spiral, which gives proteins consisting of such structures an elasticity or resilience. Such proteins are found in wool, hair, and tendons.

18.   The strands of protein in hair are cross-linked by means of disulfide linkages between cysteine amino acid units in adjacent strands. In a permanent wave, the hair is first treated to break these linkages. Then, after the hair is rolled onto curlers, the linkages are allowed to

reform. As the hair is on the curlers, the disulfide linkages form in a different place along the protein strands, which imparts a curl to the hair when the curler is removed.

20.   Oxygen is transported by the protein *hemoglobin*.

22.   Antibodies are special proteins that are synthesized in the body in response to foreign substances such as bacteria or viruses.

24.   Amino acids contain both a weak-acid and a weak-base group, and thus they can neutralize both bases and acids, respectively.

26.   The molecule acted upon by an enzyme is referred to as the enzyme's substrate. When we say that an enzyme is specific for a particular substrate, we mean that the enzyme will catalyze the reactions of that molecule and that molecule only.

28.   Figure 20.11 illustrates the lock-and-key model clearly. The lock-and-key model for enzyme action postulates that the structures of the enzyme and its substrate are complementary, such that the active site of the enzyme and the portion of the substrate molecule to be acted upon can fit together very closely. The structures of these portions of the molecules are unique to the particular enzyme-substrate pair, and they fit together much like a particular key is necessary to work in a given lock.

30.   Sugars contain an aldehyde or ketone functional group (carbonyl group), as well as several (OH groups (hydroxyl group).

32.   a.   glucose

b.   ribose

c.   ribulose

d.   galactose

34.   A polysaccharide is a large polymeric substance, containing as its building blocks repeating simple sugar (monosaccharide) monomer units. Both starch and cellulose consist of long chains of bonded *glucose* molecules: the differences in properties between starch and cellulose are derived from the different manner in which the glucose units are attached to one another to form the polysaccharide chains.

36.   ribose (aldopentose); arabinose (aldopentose); ribulose (ketopentose); glucose (aldohexose); mannose (aldohexose); galactose (aldohexose); fructose (ketohexose).

38.   A nucleotide consists of three components: a five-carbon sugar, a nitrogen-containing organic base, and a phosphate group. The phosphate and nitrogen base are each bonded to respective sites on the sugar, but are not bonded to each other.

40.   Uracil (RNA only); cytosine (DNA, RNA); thymine (DNA only); adenine (DNA, RNA); guanine (DNA, RNA)

42.   When the two strands of a DNA molecule are compared, it is found that a given base in one strand is always found paired with a particular base in the other strand. Because of the shapes and side atoms along the rings of the nitrogen bases, only certain pairs are able to approach and hydrogen-bond with each other in the double helix. Adenine is always found paired with thymine; cytosine is always found paired with guanine. When a DNA helix unwinds for replication during cell division, only the appropriate complementary bases are able to approach and bond to the nitrogen bases of each strand. For example, for a guanine-cytosine pair in the original DNA, when the two strands separate, only a new cytosine molecule can approach and bond to the original guanine, and only a new guanine molecule can approach and bond to the original cytosine.

44. Messenger RNA molecules are synthesized to be complementary to a portion (gene) of the DNA molecule in the cell, and serve as the template or pattern upon which a protein will be constructed (a particular group of nitrogen bases on *m*-RNA is able to accommodate and specify a particular amino acid in a particular location in the protein). Transfer RNA molecules are much smaller than *m*-RNA, and their structure accommodates only a single specific amino acid molecule; transfer RNA molecules "find" their specific amino acid in the cellular fluids, and bring it to *m*-RNA where it is added to the protein molecule being synthesized.

46. A triglyceride typically consists of a glycerol backbone, to which three separate fatty acid molecules are attached by ester linkages.

48. "Soaps" are the salts of long-chain organic acids ("fatty acids"), most commonly either the sodium or potassium salt. Soaps are prepared by treating a fat or oil (a triglyceride) with a strong base such as NaOH or KOK. This breaks the ester linkages in the triglyceride, releasing three fatty acid anions and glycerol.

$$\text{triglyceride} + 3\text{NaOH} \rightarrow \text{glycerol} + 3\text{Na}^+\text{soap}^-$$

50. Soaps have both a nonpolar nature (due to the long chain of the fatty acid) and an ionic nature (due to the charge on the carboxyl group). In water, soap anions form aggregates called micelles, in which the water-repelling hydrocarbon chains are oriented towards the interior of the aggregate, with the ionic, water-attracting carboxyl groups oriented towards the outside. Most dirt has a greasy nature. A soap micelle is able to interact with a grease molecule, pulling the grease molecule into the hydrocarbon interior of the micelle. When the clothing is rinsed, the micelle containing the grease is washed away. See Figures 20.22 and 20.23.

52. Cholesterol is the naturally occurring steroid from which the body synthesizes other needed steroids. As cholesterol is insoluble in water, it is thought that having too large a concentration of this substance in the bloodstream may lead to its deposition and build up on the walls of blood vessels, causing their eventual blockage.

54. Phospholipids have a similar structure to triglycerides, except that one of the fatty acids attached to the glycerol backbone is replaced by a phosphate group (phospholipids are sometimes referred to as phosphodiglycerides). An important phospholipid is lecithin, which behaves very strongly as an emulsifying agent (allowing polar and nonpolar substances to mix). Lecithin is found in high concentration in egg yolks, and is used in baking in place of whole eggs. The emulsifying properties of the egg yolk are due to the lecithin, and using lecithin itself avoids the cholesterol problem associated with whole eggs.

56. i

58. m

60. u

62. f

64. g

66. r

68. p

70. o

72. b

74. d

76. a

78. nucleotides

80. ester

82. thymine, guanine

84. transfer, messenger

86. lipids

88. unsaturated, saturated

90. ionic, nonpolar

92. fatty (long chain)

94. progesterone

96. The primary structure of a protein refers to the specific identity and ordering of amino acids in a protein's polypeptide chain. The primary structure is sometimes referred to as the protein's amino acid *sequence*.

98. tendons, bone (with mineral constituents), skin, cartilage, hair, fingernails.

100. Proteins contain both acidic (-COOH) and basic (-NH$_2$) groups in their side chains, which can neutralize both acids and bases.

102. pentoses (5 carbons); hexoses (6 carbons); trioses (3 carbons)

104.   In a strand of DNA, the phosphate group and the sugar molecule of adjacent nucleotides become bound to each other. The chain-portion of the DNA molecule, therefore, consists of alternating phosphate groups and sugar molecules. The nitrogen bases are found protruding from the side of this phosphate-sugar chain, bonded to the sugar molecules.

106.   The primary human bile acid is cholic acid, which helps to emulsify fats in the intestinal tract. In order to be digested by enzymes for absorption into the bloodstream, large clumps of fat must be dispersed into fine droplets in the liquid of the small intestine. Cholic acid basically acts as a detergent.